How Can
We Make
MANUFACTURING
Sexy?

A Mindset of Passion and Purpose from the
Production Floor to the Executive Suite

KARIN LINDNER

KARICO
PRESS

First Edition, second printing, August 2012

ISBN 978-0-9880171-0-8

Edited by: Heidi Garcia, Phyllis Mancini and Susan Chilton
            (primary editor)
Photography: Karl Schrotter
Cover and text design/layout: Kim Monteforte, www.wemakebooks.ca
Print production: Beth Crane, www.wemakebooks.ca

Printed in Canada

MOTTO

"Excellence is not a skill. It's an attitude."

Table of Contents

Preface

SPECIFICALLY FOR OPTIMISTS, BELIEVERS AND YEA-SAYERS

"Faith is passionate intuition."
~ *William Wordsworth*

Look out! This book will inspire you and increase your awareness. It was written to create excitement and passion for a new manufacturing era. These pages offer new perspectives for any person with an open mind. Answering the question *How Can We Make Manufacturing Sexy?* will help you rediscover what we are missing in the manufacturing industry because we have lost the hunger to succeed and the ability to see the possibilities. My feeling is that people who genuinely care have become the minority. People who blame, moan and whine are at the forefront. Isn't it time for a change?

I have intentionally posed many provocative questions in this book to trigger thinking and reflection on your part. The answers to these questions should come from within you. Take advantage of this opportunity to uncover your own "aha" moments. Have a pen and paper at the ready to capture them. The best way to shed light on new possibilities is to use your ability to think for yourself.

In my worldview, simplicity is key. Our complex environment has forced people to constantly cover up what they don't understand. With that, they lose a lot of their potential productivity. If we can inspire people to refuse to accept the status quo by always asking if there is a better way, there is no doubt in my mind that we will be able to stay competitive in a global marketplace.

A while ago, I was pondering why leaders such as John F. Kennedy, Martin Luther King and Gandhi were such influential speakers. They had an immense impact on the world by creating positive change for society. Why? The answer I came up with was "hope." All three powerful men were able to communicate a message of hope for a better tomorrow. From the moment I started to write this book, this has been my intent, too.

I believe we have to do a better job communicating hope in our organizations instead of presenting one stale matrix after the next. Shouldn't people be able to reach their full potential and perform to the best of their abilities? How can we help them?

We can help them by painting a clear picture of what's possible.

Preface

SPECIFICALLY FOR PESSIMISTS, DOUBTERS AND NAY-SAYERS

"Doubt is a pain too lonely to know that faith is his twin brother."

~ Khalil Gibran

Watch out! This book is going to challenge gloomy outlooks. It was written out of passion and the desire to create awareness. As a right-brain generalist, I had to conduct a lot of research and interview several people to broaden my horizons. I have to admit that I don't get excited when I hear the word "matrix." Although I believe that data and statistics are important, I would rather observe people's attitudes, behaviors and actions when it comes to business. The intent of my book is to break with past approaches and to change our mindset towards work and life in general.

I ask a lot of questions in this book but I don't always provide all of the answers. If this frustrates you a little as you start to search for new solutions to old problems, don't give up. At this point, I simply want to encourage you to get a pen and paper and discover the answers for yourself. I want you to *think* your way through this book.

When you first read the title *How Can We Make Manufacturing Sexy?*, it probably piqued your interest because you felt there was some kind of disconnect. What does manufacturing have to do with being sexy? Maybe that's the problem. We all know that sex sells and not even a pessimist can argue against this fact.

Don't get me wrong. I do appreciate pessimists (in small doses) because they have the caution and deliberateness I sometimes wish I had more of. That said, I still wouldn't want to trade places. I believe that inside every cynic is a person who actually hopes to be proven wrong.

I want to bring attention to our vital manufacturing industry. We *can* make it sexy. Read on and see how.

Chapter 1

UNCONVENTIONAL INSIGHTS OF A WOMAN IN A MAN'S WORLD

"If you think you are too small to make a difference, try sleeping in a closed room with a mosquito."

~ African Proverb

Sexy manufacturing? For some, this may sound contradictory. *In reality, every organization needs an element of sexiness.* The dictionary defines sexy as "generally attractive or interesting; exciting, appealing, trendy." It is the creation of an initial interest. You may or may not agree but from my perspective as a woman, being sexy embodies "getting attention," "demonstrating confidence" and "providing attraction," and there is absolutely nothing wrong with that. As a matter of fact, now more than ever, it is time to rethink, redefine, and thereby rediscover the sexiness of our manufacturing industry.

Ask yourself these questions:

- **Does manufacturing receive a lot of positive attention?**

- **Does society demonstrate confidence in the stability of our North American manufacturing base?**

- **Do we have strong and convincing marketing initiatives when it comes to influencing consumers to buy a North American product?**

- **Have we put enough effort into attracting the next generation and more women to an industry that is essential to our wealth and prosperity?**

- **Does our education system support apprenticeship programs and practical "hands-on" learning initiatives to ensure the development of a skilled workforce?**

If you answered no to more than one question, you agree with me: We face a serious problem.

There is proof all around us that things done with passion lead to long-term success. **Unfortunately, in far too many instances, the excitement for working in a job has been lost and passionate executives who want to create a better future with purpose and meaning are rare.** What happened? Is this the price that we have to pay for our superficial, risk-averse, indecisive and materialistic society? **Have we traded passionate leaders for passive managers?**

Doesn't it make a lot of sense that when you can make a job, product, service or business "appealing," "wanted," or even "coveted" by employees, employers and customers you have found the key to success? What's possible with the right mindset, attitude and drive is surprising. If we have sexier leadership, we will have sexier businesses and, with that, sexier employee attitudes that will result in a sexier tomorrow.

Am I clear about the motivation behind this book? I believe manufacturing leaders are "not aware that they are not aware" about the opportunities in this industry because they are too busy fighting fires on a daily basis. I want to help these leaders recognize that they actually do have the power to move from reaction to action. I want to provide them with the perspectives and tools they need to create a new manufacturing era. To make this happen, the very first thing they need is the right mindset.

I am passionate about creating positive change in the manufacturing industry and it is my firm belief that if you are not able to act proactively in changing times, your competition will easily outperform you in no time.

So who am I, a woman who came from a small village in Austria, to think that I can make a difference in the mindsets of leaders and workers in manufacturing environments? Perhaps it's because I can see things that others can't see just yet. I am passionate, ambitious, persistent, and I like to make things happen. I am also convinced that every person can make a difference if he or she sets his or her mind to it. **You can't motivate people to feel passionate, but you can inspire them to believe in and see the bigger picture and this does help bring out the best in them.**

Why would anyone be interested in reading or listening to my perspectives? Perhaps it's because there are many people who are reliant on the future of this vital sector and the wealth that comes with it. These people know that "something" needs to change but they can't get a grip on what it is, or they have challenges in making it happen. Also, I am a woman and it is a well-known fact that women think, act and react very differently than men. It's not better, it's not worse, it's just different. I fully agree with what Einstein said: "The definition of insanity is doing the same thing over and over again, expecting a different result."

Considering that most manufacturing processes and systems have been and continue to be conceived by engineers and other left-brain thinkers, it may be time for the manufacturing industry to listen to a right-brain generalist in order to gain a different perspective.

Let me backtrack a little. In May 2003, I moved from Austria to Canada. Little did I know at the time that I would discover my life's purpose at a company called Magna International, founded by a fellow Austrian, Frank Stronach. I worked at Magna until my position was eliminated due to restructuring at the end of 2006. However, working at Magna for three years is equal to working elsewhere for ten years with respect to the learning experience. To this day, I am grateful for that opportunity, which turned out to be a stepping-stone in the creation of my own business.

Thanks to Stronach, Magna is a very progressive organization and in many ways well ahead of the game. Despite that, I discovered that there were significant communication gaps between management and its workforce. I often thought about the number of missed opportunities because very few leaders knew how to tap into the brainpower of their employees. They, no doubt, would if they knew how. This is something that can be easily learned but requires an open mind and the willingness to do things differently.

I further discovered that people seem to get more enjoyment from the "blame game" than from looking for creative solutions for the benefit of the team and the organization. People expect others to change. Very few are willing to focus on themselves to the same end. Actually, the higher their rank, the worse it becomes.

We judge ourselves by our intentions and others by their actions. That's simply human nature.

If I came to these unconventional insights at a progressive organization such as Magna, you can only imagine how much untapped potential there must be in other organizations. That was my "aha" moment and it was then that I realized I had found a niche market for my business. From my own experience, I know that Citizenship and Immigration Canada is endeavoring to attract well-educated immigrants into the country. Could it be that our corporate leaders consciously choose not to take advantage of our diverse workforce and overlook this valuable competitive edge? I don't believe that this is as much a conscious choice as attributable to the fact that most of them don't have the time, or don't take the time, to even consider it. They can't see the forest for the trees.

A lot of my thinking about what's needed in manufacturing today comes down to authentic leadership. I also believe the workforce has to make a shift in its mindset. There must be give and take on both sides. People deserve to like their jobs but if they are willing to collect a paycheck, they shouldn't have a "can't do" or "minimum required" attitude. **An attitude of entitlement seems to have pervaded North America and we don't have to look far for examples. It's a self-inflicted misery that holds businesses back from performing at their best.**

In his research on motivation, William James of Harvard University found that hourly employees could maintain their jobs (that is, not be fired) by working at approximately 20 to 30 percent of their ability. His study also showed that highly motivated employees work at close to 80 to 90 percent of their abilities. Can you imagine the lost potential? Something clearly has to change. In a new manufacturing era, people must learn to show initiative, responsibility and ownership. What is being done to create and develop a "skilled workforce" with the drive and initiative to produce higher quality products? This would certainly be one way to differentiate ourselves from lower cost countries.

There was a time when having a "factory job" was preferred over many other careers. Manufacturing and trade creates wealth — even if that wealth is later traded for services. As an advisor on employee engagement and motivation, I would really like to see more homegrown accomplishments in manufacturing and less offshoring to lower cost countries. Wouldn't you?

Competitors such as China, India and Eastern Europe are yearning to succeed and that's what we have to re-discover: the drive and the desire to succeed. We cannot accept being a society known for its greed, complacency and sense of entitlement. Instead, we must understand that we live in a constantly changing world and we must adjust. The sooner we do, the better off we'll be.

Today's manufacturing leaders can no longer depend on the answers to yesterday's challenges because the dynamics we face in this fast-paced, global market are new. This book is based on my personal definition of the "rule of thumb," but for me T.H.U.M.B. stands for Thoughts, Habits, Understanding Emotions, Mindset, and Beliefs. I feel strongly that using T.H.U.M.B. can have a significant impact in this new global market by helping people increase their awareness of these factors and by positively influencing their actions and reactions in business. This should awaken in them the desire to reach the next level.

So what are we waiting for?

Don't you think it would be worth the effort to put our brilliant minds together to make it happen? Just imagine the benefits to our organizations, our workforce, and our overall society with a creation of newfound wealth and wellness!

If you agree that it is time to consider personalizing and pumping up business by combining facts with feelings, keep an open mind while we explore the many opportunities that are within our reach in the manufacturing sector.

Chapter 2

HOW DID WE GET INTO THIS MESS?

> "Money doesn't change men, it merely
> unmasks them. If a man is naturally selfish
> or arrogant or greedy, the money
> brings that out, that's all."
>
> ~ *Henry Ford*

What mess am I talking about? This may be a loaded question, but I have noticed that many people are in denial about what's going on.

You may agree that it's much easier to close your eyes, hide behind excuses, blame somebody else, or simply pretend that everything is going well.

I could talk about many things in this chapter. I could talk about unions, the government, labor rates, corporate tax rates, federal regulations, and ever-increasing gas prices. However, I decided not to write about these topics but instead focus on things that are well within our control.

Patience is a virtue, but unfortunately it's not one of mine. **If you are looking toward the government or your local politicians for solutions, you are looking in the wrong place.** We cannot wait for the government to take action. Time is of the essence. The challenges facing the manufacturing sector cannot be driven by lawyers, bankers or any political group. **This is an issue that must be driven by passionate people within the industry** and organizations such as the Society of Manufacturing Engineers (SME), the Association for Manufacturing Excellence (AME), the Canadian Manufacturers & Exporters (CME), and the Automotive Parts Manufacturing Association (APMA), to name just a few. Together we can direct the focus on education and the creation of new talent that is vital for continued innovation and proprietary manufacturing. Perhaps we can attract enough attention that politicians will have to sit up, take note and get involved.

The entire focus of this book is on what our manufacturing leaders and employees can do right now to make a difference and to have a positive impact on this industry. I am not interested in pointing fingers at the government, the legal system, our culture, exchange rates, or the cost to bring to market. While it may be human nature, it's time to move away from a blaming mentality in order to see what we can change. Of course we have to be realistic, but I believe in overcoming obstacles and in the power of "making it happen." The history of the United States and Canada always has been and always will be about overcoming obstacles. If we take the initiative and start building something new, others within our circle of influence will follow. **I am certain that we can create a positive movement of forward momentum.** A society of whiners won't be able to move forward, but a society of winners can and will change the mindset, drive, dynamic and vision of the future. Fortunately, we can inspire our leadership teams to embrace the power of imagination.

The automotive industry has played a critical role in the expansion of America's middle class for nearly a century. Yet the industry has suffered great losses in recent decades and was deeply affected by the recession of 2008-09. Auto sales fell by 40 percent, and two of the three automakers in Detroit filed for bankruptcy protection.

Do you recall the day in 2008 when the CEOs of the Big Three flew to Washington in private jets to ask the government for bailouts, utilizing taxpayers' money? They took a lot of criticism for that and rightly so. How often have we read about similar instances in the newspapers? We become outraged or upset and then we forget about it...until the next time we read something similar.

- **Why is there always a next time?**

- **What has happened to our integrity?**

- **What has happened to our business ethics?**

- **What has happened to our drive, ambition and hunger to be innovative and creative?**

- **What has happened to our corporate leaders as a whole?**

- **Are we running out of hope?**

- **Do we have to wait for the next generation, hoping they've been raised to not only believe in values but also to live them?**

If this is the case, I hope that our business schools and universities recognize the importance of teaching students about the power of attitude, behavior and a positive mindset in order to succeed in this fast-paced, global marketplace.

So, how did we get into this mess? Why do we have to ask ourselves questions like the above?

The simple answer is that it was due to a lack of leadership, too much greed, and no imagination about what could be. Our leaders at the top are completely disconnected to the people who are supposed to get their high-quality products out of the door. If instead of having a give and take approach, the leader just takes, of course the workforce will not care about the company. So the real question is this: "How can we change this mindset and start to lead?"

Following a period of prosperity, 2009 was a wake-up call. It was a scary time for many of us. **Now more than ever, I am convinced that we were faced with more of a leadership crisis than an economic crisis.**

It is my observation that the best way to take passionate and self-motivated behavior out of business is to replace it with obscene amounts of money being paid in the form of ridiculously high salaries, bonuses and discretionary rewards. I am the first one to agree that if executives do an outstanding job, they should be well compensated for their high-stress and time-consuming positions, but how much is enough?

I have given this a lot of consideration, and would like to ask you this: Don't you think it's about time we take a look at the golden handshake? How can it be that an executive is allowed to run a company into the ground and still be rewarded with millions? Does that make sense to you? Enough, already!

It's 2012; isn't it time to rethink these extreme executive compensation packages? What motivates senior-level executives to be ethical and lead organizations with high integrity and sound judgment if they know that they will be rewarded regardless of whether they succeed or fail? What about corporate governance, boards of directors, and compensation committees? How can they not unequivocally specify that an executive must perform well in order to receive bonuses, stock options and so on? All of us know of some high-profile instances where executives cashed in big time despite the fact that under their stewardship their companies lost millions of dollars and thousands of workers were laid off. How can this be justified?

My perspective on this topic was further impacted when I read Daniel H. Pink's book *Drive: The Surprising Truth About*

What Motivates Us. This book is based on scientific research and because it has a very compelling message, I want to share some of Daniel's insights on human motivation.

It discusses the idea that "for as long as any of us can remember, our thoughts were based on one assumption: The way to improve performance and increase productivity is to reward the good and to punish the bad." This simple axiom still serves some purposes well; it's just deeply unreliable. Sometimes it works, but often it doesn't.

When people aren't producing, companies typically resort to rewards or punishment. They try to avoid doing the hard work of diagnosing what the problem or root cause actually is. It's easier to utilize the carrot or stick methodology.

"The 'if-then' motivators often stifle, rather than stir, creative thinking." But could that really be? If you want people to do better, you reward them with incentives, bonuses and commissions. Pink suggests that, "Rewards in and of themselves dull thinking and block creativity." For any task that requires real problem-solving abilities and possibility thinking, monetary rewards won't work because they narrow the focus and concentrate the mind on only one destination. He says, "There is a mismatch between what science knows and what business does."

Various experiments have shown that bonuses work as long as the task involves only mechanical skill — the higher the pay, the better the performance. However, higher incentives lead to poorer performance when it requires cognitive skills and creativity.

The approach we should take is very much about intrinsic motivation, based on the desire to do something because it matters, because we like it, because it's interesting, because it's part of something bigger.

The solution to getting out of this economic mess is not to do more of the wrong things.

*After reflecting on all of the insights from Pink's book, I am
left with some burning questions:*

- If what science suggests is true — that financial incentives can negatively impact overall performance — how does this apply to executive teams?

- Is what they do mechanical or cognitive?

- Should we seriously consider eliminating big bonuses to bring back real purpose, passion and meaning?

- Isn't it true that quarterly earnings have become an obsession for many leaders? In order to cash their bonuses, would managers rather take the quick road, even if it means the low road, oftentimes not contemplating the consequences or long-term health of the company?

- Is it acceptable that our society associates prestige with how much money we make rather than the class and character we display?

- Shouldn't we stop making it all about the money and "me, myself and I," and craft a new and better operating system to help society and companies achieve higher levels of excellence?

We need to realize that to survive, we must compete in a global manufacturing economy. We have to get over our North American sense of entitlement. We have to overcome our sense of hopelessness. We have to keep working at it, getting better plant by plant and worker by worker. That's how winners stay — or get back — on top.

Susan Helper is a professor of economics at Case Western Reserve University in Cleveland. She and her team conducted an interesting study on workforce change. Her research aligns with my observational findings in the industry. **"Everything is short** term. Today is everything. That mentality drives the behavior: Get it now, and don't worry about the out years... Focus on today; worry about tomorrow, tomorrow. When companies expend energy on discrediting one another instead of working to address the root causes of future problems, the immediate result may be profitable for one firm or another, but the eventual pattern of behavior is both wasteful and debilitating to the broader network."

Professor Helper has studied the auto industry for over two decades. She frequently appears as an industry expert in newspapers such as the Wall Street Journal and the New York Times. She is also a judge for the Automotive News PACE Award for supplier innovation. Her extensive research can be found at: http://drivingworkforcechange.org/reports/supplychain.pdf.

Being European myself, I would like to point out some of the reasons behind Germany's manufacturing and economic success and why I believe North America has some catching up to do:

- Germans start cultivating an interest in prospective manufacturing and engineering in kindergarten.

- The manufacturing programs taught in their universities utilize mock factory floor layouts to teach undergraduates the ins and outs of factory life and problem solving.

- Apprenticeship programs are widely promoted and considered a must-have. The guidance counselors in high schools do not send 25 percent of the high school population to colleges or universities. More realistically, they encourage many students to attend apprenticeship programs to develop high-level manufacturing skills.

- There are two economies that are weathering the super-recession well because they produce and export goods. The first one is China, long competing on low

wages, cheap goods and exchange rate advantages. While the wage gap is declining, China will remain competitive due to more and more other factors. As an export-focused economy, the second one doing well is Germany because of its specialized and highly engineered premium products. German engineering is known for its high quality standards.

The question now is what route will North America choose?

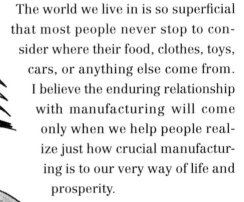

The world we live in is so superficial that most people never stop to consider where their food, clothes, toys, cars, or anything else come from. I believe the enduring relationship with manufacturing will come only when we help people realize just how crucial manufacturing is to our very way of life and prosperity.

Chapter 3

HAVE YOU MET
YOURSELF LATELY?

**"Everyone thinks of changing the world,
but no one thinks of changing himself."**

~ Leo Tolstoy

This is one of my favorite quotes. We all want change. We want other people to change. We want our environment to change. We want circumstances to change. Most manufacturing management teams look for the silver bullet to create immediate change, but do they really *want* to change? Do *you* really want to change?

I am convinced that change requires many things, but it starts with courage and commitment and, from my perspective, very few business leaders today have what it takes to stay ahead of the game.

Our business environment is not only changing, it is going through a major transformation. We are confronted with issues that we have never dealt with before. Young people today expect to be led in very different ways than previous generations, and pretty much everything that used to work with customers and suppliers no longer applies. Business today is daunting, demanding and highly competitive, but it can also be a lot of fun. It is another new challenge for North Americans to overcome. Shift happens! Our high-tech environment makes change inevitable and we have to adapt in order to survive.

It is unfortunate to see examples of disengaged individuals — managers, supervisors and employees — who seem to have settled for mediocrity as a result of being too comfortable where they are. In many workplaces, disengaged leadership styles abound and it is important to understand that a disengaged manager will never have an engaged workforce.

The dictatorial manager is one example of a disengaged leadership style that is still alive and well in too many companies. Employees simply do what they are told to do, no questions asked. Forget creativity and innovation. Employees are merely seen as bodies, not minds.

Why would any business owner want to have employees who have mentally left the building? My simple conclusion is that they probably don't realize it and if they do, they simply don't know how to mentally re-engage them. So my solution would be to create a willingness to radically improve certain leadership behaviors in order to re-ignite employees. Impossible? **Believe me, big changes can be created quickly and with small steps.** Where there is a will, there is a way! I have seen people rediscover their inner drive to take on new challenges. It's just a matter of leading the way.

I have noticed that many men and women have already become much more selective when it comes to choosing future employers. They carefully observe how they are treated during the interview process instead of being fearful and accepting just any job that is offered. They research the reputation of the industry and they ask around before they even apply at a specific company. Employees who do their homework and research to ensure a good job fit are exactly the kind of exceptional, value-driven people you should aspire to attract.

During the hiring process, companies have to change their mentality about being in a position of power just because they are offering a job. They should be more concerned about what kind of a first impression they make on potential employees who are supposed to contribute to the company's growth and success. What etiquette does your company follow when a candidate arrives at your door? Do you welcome him or her on time? Do you offer a refreshment? Do you show an interest in his or her personal interests? Is the room where the interview will take place neat and tidy? Have you prepared for the encounter?

On the other hand, I have also noticed that many employees feel a sense of entitlement and have grown accustomed to be taken care of. **A cordial greeting or a polite "please" and "thank you" are no longer part of our daily routine at work.** Employees can frequently put forth great effort to ignore one another rather than going out of their way to help. A common workplace issue today is that few employees show initiative, and fewer still take full ownership or responsibility. Bad attitudes, an overall lack of professionalism, and high levels of negativity are commonplace.

All of this seems to be widely accepted or unnoticed by management. Very few are held accountable for unacceptable behaviors. That's usually when I'm called in. "Karin, can you please help us to get our employees more motivated and engaged?" Many times I enter organizations and somehow executive teams seem to think that I sprinkle some pixie dust to "fix" their people. A quick motivational seminar should do it. I'm afraid not, yet they often add insult to injury by asking what that would cost instead of what the outcome would be. I wish I could wave a magic wand, but there is no such thing.

Other corporate leaders have asked me how they can change the thinking and attitude of people in their department or how they can get them to adjust to new situations. My answer is always the same: As a leader you cannot change people but YOU can keep learning and growing in order to help them want to make changes. If you want things to change, you've got to change. If you want the work environment to get better, you've got to get better.

The good news, I tell them, is that I can help. I can ignite the creativity, energy and spirit in the employees. I can also help reduce costs and save money fairly fast but it will be sustainable only if the company leaders are willing to follow my common-sense advice. **Anything is possible if a leader is willing to develop a "grow together with me" attitude.** Sustainable operational improvements cannot be achieved without touching people's hearts, spirits and inner drive to become learners.

The long-term goal has to be sustainability and this requires trust, courage and commitment from everyone in the organization. No exceptions! If I have learned one thing over the past few years it is that it is a waste of time for me and money for them if I'm making commitments to uncommitted people. Commitment is a word most people shy away from because committing to something means actually doing it. It is often easier to remain vague.

Have you ever considered that a team is a reflection of its leader? I know that it is difficult to take a critical look at ourselves because we are all legends in our own minds. How do I know this? I know this because I too am a legend in my own mind more often than I would like to admit. However, when that happens, I have to ground myself, reflect and re-focus. I know how difficult it is to be a good person and consistently concentrate on self-development in order to be a worthy leader. It is not always easy to walk the talk. If you expect more from others than you do of yourself, it is highly unlikely that you will ever be well respected.

Think about this: We spend the first 20 years of our life learning. We learn literally everything from walking and talking to studying and life skills. We're like sponges, soaking up new knowledge and talents. Then what happens?

Guess when many people stop their quest for continuous learning and self-improvement? That's right: when they land their first job. After they finish school or get their university degree, they think, "Thank goodness that's over!" and all they want is a job. The learning curve slows down or comes to an abrupt halt.

Sadly, this means that many people do not advance a lot in the next 20 years. They wrongly think they have arrived by obtaining a specific position.

This leads me to seriously question our educational system because school is where we start to lose our natural curiosity. Suddenly the teacher is asking the questions and the students have to have the answers to show how smart they are. We unlearn posing questions and become conditioned to finding the right answers. This presents a real danger because when people enter the corporate world, they believe they've achieved their end goal and they can stop learning. Do you sometimes feel overwhelmed thinking that you have to have all the answers and that you ought to be beyond asking the questions? How can we create more openness when it comes to learning from the people around us? How can we become more open-minded when it comes to working on ourselves?

I think that one of the most difficult things to keep in check is our ego. I make a conscious effort every day to check my ego at the door when I enter the offices of my clients. This certainly is not always easy. However, I remind myself that my goal is to serve them and bring value to their lives and to their businesses. Whether I'm a small-group facilitator or speaker in front of a large group of people, I keep in mind that it is not about me; it is all about them, their needs and bringing out the best in them. It is about showing a genuine interest in other individuals and letting them know that you care. **In the end, if you don't care about your employees, why would they care about your business?**

Unfortunately, we live in a society that is based on a "me, myself and I" attitude. Isn't it true that there are too many self-serving people in most organizations? I know many companies where the top executives tell their employees to save money or to cut costs but somehow the same rules don't apply to them. They still cash in big bonuses, fly business class, spend money on expensive dinners, and are puzzled when the number of disengaged employees increases and morale sinks lower and lower.

What has happened to following through on a long-term vision? What has happened to taking a step back and taking one for the team?

If you are interested in conquering your ego, here are some questions for you to entertain:

- **Do you sometimes think that there is nothing left for you to learn?**

- **Do you have the need to tell everyone how smart you are or boast about your accomplishments?**

- **How easily can you express regret?**

- **How well can you accept constructive criticism?**

- **When you make a business-related decision, do you have the company's long-term vision in mind or do you go for the immediate financial gains?**

- **Do you play by the same rules as everyone else or are you the exception?**

- **Do you feel that because of your position you have the right to make negative or derogatory comments to the people around you?**

It is my goal to inspire extraordinary leadership at all levels of the organization in order to create positive change in manufacturing environments. **I strongly believe that when you are able to demonstrate courage, you can encourage people to try new things. With that, you definitely can get to the next level.**

Chapter 4

YOU GET WHAT
YOU EXPECT

"The way we see the problem is the problem."
~ *Stephen R. Covey*

Have you ever noticed that people spend a lot of time focusing on problems? They look for problems all day long and 99.9 percent of the time they get just what they asked for. It seems to me that even new and highly motivated employees are getting worn out by this mindset.

Recently, I encouraged a group of workshop participants to brainstorm various ideas in response to a very simple question. One person demanded, "Don't we have to define the problem before we start brainstorming?"

We have become so "problem-centric" that we can't see all of the opportunities around us. It's as if we are blindfolded. Unfortunately, I do not come across too many gung-ho enthusiasts who are willing to take a leap of faith for innovation and creativity in manufacturing environments.

Is this the reason that we tend to move sideways or in circles instead of forward? Why is it that some people think there is only one way to approach a problem? I have so often seen teams clinging to the same solutions they had ten years ago. I have said it many times before and I will say it again: Management can no longer depend on old answers because the problems companies now face are entirely new.

I enjoy analogies and the crab seems to be a good one. Crabs move sideways instead of forward, but do you know why? A crab has evolved to move sideways in response to its environment. Its body is relatively flat with eyes that face forward on stalks. It lives in rock crevices and moves sideways across the surfaces of coral and sand with its back against the rock face. It's forever facing outward to protect itself from possible threats from predators and other crabs. When danger appears, it can easily conceal itself in gaps between rocks for protection.

Sometimes I find this behavior in the manufacturing industry. What kind of work environments have we created? Do you feel as though your back is against the wall? Are you anticipating problems and responding to customer demands all day long? Are you in a "crab-like" mode?

Many manufacturing companies in North America hold on to the "same old, same old" strategy. We have become problem focused instead of solution oriented.

The author Anais Nin wrote: "We don't see things as they are, we see them as we are." Two people may be faced with precisely the same problem. One does nothing but complain while the other sees the opportunity in the problem, viewing it as a challenge or a chance to learn something or to make something better. This person will take action.

Do you know more people in the first category or the second?

What about you? Do you see problems or opportunities?

Have you ever considered that the only genuine problem you may have is your perspective? Could there be something amiss in how you look at things; how you perceive everything that happens, not only at work but also in your personal life?

It is not difficult to spot problems. We all excel at that. As frustrating as it is, we like to complain but rarely do we take the initiative to do something about it. We prefer to let someone else deal with it. Think of how many times you have heard someone say, "That's not my problem." The question is, "How can we change our perception of problems?"

I often suspect that the more people talk about their daily trials and tribulations, the more important they feel. It's like they are screaming to the world, "Look at all the nasty stuff I have to deal with. Poor me!" Why do they do this? It's because the majority of us are not being validated in a positive way, but when we talk about all the "bad stuff" we endure, we get attention. That's one of the main reasons that people never stop talking about what's wrong and rarely acknowledge what's right.

There is no doubt in my mind that the best way to solve a problem is to change how we talk and think about it. For example, let's call a problem a roadblock. Problems are obstacles and obstacles need to be overcome with determination, willpower and strength. We need to change our mindset, our perception and our attitude when it comes to dealing with unexpected, challenging circumstances because in essence, they make us stronger, test us and keep us on our toes.

Let's recall for a moment how we feel when we talk about problems. We may feel frustrated, discouraged and often overwhelmed because we don't know how to deal with the situation. Our energy level decreases. It sucks the life out of us.

Now imagine rephrasing a problem as being an opportunity or possibility. How do we feel when we talk about new opportunities in the market? How do we feel when we communicate about the different possibilities in our industry? Our energy level immediately increases. Can you see the excitement that you could create within your team just by changing your language and how powerful this could be?

Have you ever wondered why things always seem to work out

for some people and not for others? Have you ever thought about why some people never get past the stage of trying? I once heard a saying that I believe to be true: "Some people are not aware that they are not aware that they are not aware." Our mindset and attitude can undoubtedly hold us back from moving forward.

Did you know that the average person has about 60,000 thoughts a day and most of these thoughts tend to be negative? In the stage of unawareness, these thoughts can turn into a powerful voice that we listen to all day long. It is human nature that we tend to put ourselves down. "I am not good enough, I am not smart enough, I am not worthy enough, I am not good looking enough, I cannot get this job done," and the list goes on and on. How aware are you of the self-talk that's going on in your head? Many of these thoughts are voices from the past, created from earlier experiences and stories people have told us. If all of these experiences and stories are powerful enough or told to us by a trusted source, they can become ingrained and form a part of our belief system. When we have to make certain decisions, we may still hear the echoes of our father, mother or teacher.

What about you? Do you have more positive or more negative thoughts?

You have to be careful because this critical little voice is trying to protect you from fear of failure and rejection in a highly unproductive way. Negative thinking may become automatic, embedded in your self-image and you end up living down to its expectations. I have two simple words of advice: Stop it! You don't want that niggling insecurity for yourself or for your organization.

I consider myself to be a very positive person and yet I still have to struggle with this turbulent troublemaker in my head. I believe the only advantage I have is that I now know how to manage my negative self-talk because I am fully aware of it.

Here are some things you can immediately start doing so that you, too, can become more self-aware:

- Consciously tune into your thoughts and work on your awareness.

- When you catch yourself thinking negatively, say STOP out loud and this will interrupt your thought process immediately. (Just be careful not to blurt it out in public or people may wonder what's wrong with you.)

- A quieter method is to put a rubber band around your wrist and snap it whenever you put yourself down. This is a neuro-linguistic programming technique, an approach to psychotherapy and organizational change that is very effective. At the beginning, you may have to deal with the side effect of having a rather red wrist. But believe me, this method works. It can help you literally "snap out of it."

- Last but not least, practice positive self-talk. Positive affirmations are one of the best ways to reprogram your brain. Sound weird? Try it! Your brain is like a computer and the input is whatever you choose it to be.

If I asked you if you have something in your life right now that started with a simple thought, what would you tell me? Yes? Of course you would. Everything we think about and focus on, we tend to achieve. If we don't, it was not important enough.

In my opinion, positive self-talk should be communicated to children in schools, in every business school, at all universities, and in every business environment. The "law of attraction," a century-old idea that is widely embraced today, refers to the idea that thoughts influence chance and results. The law argues that thoughts (both conscious and unconscious) can affect things outside the mind, not just through motivation, but by other means.

Like attracts like, if you will. In my experience, many business owners make themselves familiar with this technique and the successful ones have practiced it on a consistent basis.

The law of attraction would mean that the more our minds are focused on negative things, the more we think about problems and missed opportunities, the more challenges will come our way. How do I know? Well, I had to learn that myself. The more I resisted certain obstacles that came my way, the more difficult my situation seemed to become. The more you resist, the more it persists. Does this ring true when you think about your own circumstances at work each day?

I am not saying that I envision or even want a future without problems, obstacles and roadblocks because they are there for us to learn and to grow. However, I do believe that we have to make a conscious effort to start changing how we think and how we do things.

Once we start speaking and communicating in a more positive and encouraging manner, we will be able to create a completely different energy in our work environments. Can you see how powerful this concept could be when applied to the manufacturing industry? If you can instill in your employees the belief that anything is possible, guess what? Anything will be possible and the sky will be the limit.

Chapter 5

WHAT SHOULD MANUFACTURERS STOP DOING?

**"If you stay in this world, you will
never learn another one."**

~ *W. Edwards Deming*

As we discussed earlier, history repeats itself. Why is it that we never seem to learn from the mistakes that have been made in the past?

For heaven's sake, please stop outsourcing work overseas right now! Every manufacturer or supplier who outsources products that could be manufactured in North America needs to seriously understand the negative impact this is having on our communities and on our economic buying power. We have to stop the corporate paradigm of wanting to outsource products from each and every North American plant. Shouldn't we become more thoughtful, selective and socially responsible when it comes to producing our products somewhere else? We should consider what makes sense and what doesn't. We certainly have to stop the economic shift, which is putting far more emphasis on becoming a service-oriented industry, and go back to our grass roots of manufacturing because society's wealth, prosperity and well-being are created on goods made in one's own country. Many products that we sell here could also be made here! The devastating truth is that even though many manufacturing companies are profitable, they choose to outsource for yet higher profitability when they could produce locally. In many cases, greed is the ugly motivator and in the long run, companies will pay a hefty price for that. Have people given up on the idea that we can be a self-sustaining country and as a result chosen to take the road of least resistance?

It is extremely and undeniably difficult to tackle challenges that we have never experienced. While it is hard initially, it will become easier once we roll up our shirtsleeves and figure out what to do. **What's important is how we approach, act upon and respond to challenges.**

Do you sometimes feel that you are stuck repeating old behaviors?

Do you feel something has to be done in a different way, but you are not sure what it is?

I notice that many organizations are stuck in the past. Some are in denial and don't want to change and others want to change but don't know how. I interviewed many people and heard a lot about what manufacturers should stop doing. In fact, one of the oft-repeated comments was that industry leaders should stop searching for the silver bullet when it comes to the creation of sustainable success. It is clearly not going to be that easy.

Why do so many companies look for the quick fix?

Why is short-term gratification more important than viable long-term solutions?

Why are band-aid solutions so popular even though they never lead to enduring success?

Part of the reason may be that some leaders start with their exit strategy in mind. Some people in management don't see the benefits of working on long-term solutions because they may not work for the company long enough to reap the rewards. This may be true but it is the leaders' loss if they don't take their roles seriously enough every day that they're there to become difference makers in the eyes of their employees.

There are leaders who may not be aware of or who have never considered the positive impact they can have on the lives of their employees and on the long-term success of the organization. There is nothing more rewarding and meaningful than seeing an organization blossom due to the commitment and dedication of a progressive and caring leader. It's rewarding for the team, it's rewarding for the community, and it is also rewarding for the leader.

One of the most common phrases I hear in manufacturing environments is "I know that," and that's exactly the kind of attitude that we need to change in order to win against the competition. Knowing does not equal doing, wouldn't you agree? There is no question that our minds are filled with many things — some are useful, some are not. Hopefully our intuitive wisdom will serve us well, but holding on to old beliefs that no longer apply can prevent us from moving to the next level.

I sometimes argue that too much knowledge prevents us from learning something new. My dad says, "The more formal education people have, the less room there is for common sense." There is only so much space in our heads and sometimes we have to let go of old beliefs in order to make room for something new. How great would it be if we were open-minded enough for a fresh start? No preconceived ideas, no old belief systems, less intellectual knowledge, just the ability to come from a place of genuine curiosity. Now that would give us a new perspective!

If more people were aware of their mood or frame of mind when they left for work, it would already be a step forward. Many people are stressed, frustrated and overwhelmed before they even arrive at the plant or office. How can this negative frame of mind ever lead to a positive and productive day? I have heard people say that happiness at the workplace is overrated and in many companies happy people are even perceived as not being busy enough. Nothing can be further from the truth. This is a misconception that needs to change. Happy employees are fulfilled at work and fulfilled people show initiative, responsibility and ownership in their jobs.

Why would we want to give staff and colleagues the feeling that work has to be absolutely serious and that there is no fun allowed? Drudgery may have been an accepted practice in the industrial age, but in this age fun, passion, meaning and purpose will be essential elements to achieving results. Manufacturers who want to surpass their competition and be successful in the future need to be mindful of this fact.

There is no doubt that manufacturing can be trendy and can attract high-caliber candidates if we can stop the old-school management mentality. We need to think about ways to bring out the best in people.

- **Why are you in the position you are in?**

- **How are you going to benefit your people and, ultimately, the organization you work for?**

- **Isn't the best way to lead a group of people to work on becoming a better person and to lead by example?**

- **How can we improve ourselves to inspire others?**

Most manufacturing companies share the "tell-and-do philosophy." They tell people what to do and how to do it, instead of giving them the opportunity to find their own solutions. This leaves people with no freedom to experience any kind of self-management. We are not good at sharing our wisdom and knowledge with each other and I am not sure why corporate cultures create this immense insecurity in people. The thought process from management and supervisors trickles down to the employees and everyone shares the same belief: "If I tell you what I know, you may become smarter than I am and take my job." Why don't we establish self-confidence in people and show them the many benefits of sharing their knowledge with others? Why don't we reward information sharing?

I have always believed that it benefits me when I surround myself with people who are smarter than I am. Why? It raises the bar. It forces me to get better and I can always learn something from them. When I facilitate my workshops, I consciously avoid having the mindset that I am "the teacher" because every person brings knowledge and experience into the room. Never does a day go by when I don't learn something from my workshop participants. It is very dangerous to think that we know everything and that we have nothing left to learn. If ever you think that you are smarter than everyone around you, think again!

Begin to see the value that other people bring to the table. "It's my way or the highway" is out and can no longer be upheld. **I can guarantee you that the brain of a high-level executive and the brain of a worker is the same size.** There is absolutely no difference and the things we can learn from the people in our immediate environment can be of incredible value if we allow ourselves to see that.

When I work with manufacturing companies, I always observe the energy level in the plant. I have visited companies where I have seen employees walking around like zombies. I find it disturbing that nobody seems to notice that working without thinking cannot possibly benefit the organization. Is it a lack of awareness or the fact that managers spend too much time in their offices to "see" what's going on? I also find it kind of freaky that there are still organizations who want humans to act like robots while on the other hand they want robots to be as smart as humans.

Let's face it: Whatever problem we may have, we think we have the answer. We do what we think is the right thing to do.

We believe that because of our education and our knowledge, we are the best equipped to deal with it.

- **Have you ever asked someone a question without really listening to or acknowledging the answer because you already had an answer in your mind?**

- **What if management would learn fewer hands-on management styles and let their employees manage their own world?**

- **What if supervisors could come to appreciate the knowledge of their teams and actually consider that they may have brilliant ideas instead of feeling threatened?**

- **What if department managers would give their teams permission to continuously seek out a better way?**

- **What if plant managers would trust their employees to work as self-directed teams?**

- **What if CEOs would see it as their sole purpose to increase the self-worth and self-esteem of their teams?**

- **What would it be like if management would fight fewer fires because all the decisions they make would factor in long-term benefits rather than immediate gratification?**

I believe the smartest people are not only the ones who ask the questions but who are also ready, open-minded and willing to hear the answers.

Unfortunately, over years the belief has been established in many of us that instead of understanding the world, we have to show the world how intelligent we are. Very few of us know how to ask questions and feel confident enough to truly listen to the responses. How can you be a manager and not know everything? You have to constantly show what you know, don't you? In many minds, asking questions would mean that they don't have enough knowledge to fulfill their duty of being a great, "go-to" manager.

I cannot help but wonder what our manufacturing environments would be like if leaders would start asking more questions instead of trying to have all of the answers.

Chapter 6

HAVE THE COURAGE
TO ROCK THE BOAT

*"If you believe you can, you probably can.
If you believe you won't, you most assuredly won't.
Belief is the ignition switch that gets you
off the launching pad."*

~ *Denis Waitley*

Business transformation is exciting and we are in the midst of writing history. However, what's exciting for some may be stressful for others. My goal is to increase awareness that we have to look at the world from a different perspective. It is no longer "business as usual." It is difficult for me to tell successful companies that their success will dwindle if they continue to accept the status quo, if they keep doing what they are doing, and if they are not flexible and open to change. **My grandmother always used to say, "It is not difficult to clean the house, but it is very difficult to keep it clean."**

Within a business context this means that once you have achieved a certain level of success, it is too easy to lean back, become complacent, and enjoy the fruits of your labor. However, if you are not prepared to continuously look for better ways of doing things, you will not be able to stay ahead of the game. Success can be fleeting because it is tempting to rest on your laurels when everything seems to be going your way.

We have seen it all before when we reflect back on those large, successful corporations who suddenly got into deep financial trouble, and we will, no doubt, see it again. Arrogance and denial can be strong contenders against being flexible and proactive in a changing market.

When I think about my experience as a marathon runner, it's always easier to lead the pack than to catch up to a crowd that has left you behind. In business you don't want to be left behind, either.

Just as in an exercise routine, flexibility is the ability to stretch and that's not always easy. Although it may not feel comfortable, it is important that we stretch ourselves out of our comfort zones. **It takes hard work and commitment to stretch your body and it takes commitment and courage to stretch your mind.**

I am not saying that you shouldn't stay committed to your decisions, but it is very important to be flexible in your approach. From my own experience I know that it can be enticing to do what I have always done because if it worked before, why shouldn't it work again? However, I have learned to always ask if there is a better way.

I often hear phrases like these in my workshops: "I know that," "We have always done it this way," "I told them how to do it," or, "We don't have the time to try something new that might not work." I guess this is our "get it right the first time without making any mistakes" mentality, but I can vouch for the value in making mistakes. **Mistakes teach us to get better and better as we move along. Children advance the most quickly because they also make the most mistakes.**

If we stretch our minds far enough to look at a challenge from a different angle, it may initially be uncomfortable but on the bright side, we have another perspective that will help us generate new and exciting ideas. I am convinced that if you want to be one step ahead of your competition, you must become a flexible thinker. **You and I both know how much emphasis there is on innovation and creativity, but how can we be innovative and creative if we are not willing to work at having flexible minds and if we don't allow employees to use their brains and exercise this flexibility?**

The overall trend still seems to be more reactive than proactive and it is tough to stay ahead of the competition with this attitude. The playing field and rules have changed and we are doing business in a different world. Now, more than ever, it is important to create a positive and dynamic outlook for the future. We are surrounded by too much negativity and it is vital to change the way we think, learn and grow. Reaching the next level means being creative and innovative in order to outperform the competition.

Furthermore, most people are reluctant to admit that business involves a lot of emotion, and they very rarely make room for it. If your company has had to deal with recent layoffs, this is a definite concern. Emotions need to be dealt with, processed and released — otherwise, they can and will hurt your business.

Let's compare your workforce to an iceberg. As you know, an iceberg is a large piece of ice floating in open water. Typically, only one-ninth of its volume is visible. It's the same with your workforce. You see only a highly superficial part of them and the rest is obscured from your view.

Of course, you may not want to talk about that. You are a professional and you want to keep everything on a professional level. Well, that's exactly why things don't go the way you expect them to go. Keep in mind that if you know the "why," the "how" becomes easy. If you don't deal with what's going on inside people, it won't translate to their outside behaviors. **In other words, if you try to ignore their emotions, feelings and beliefs, you won't see the expected outcomes.**

All of the difficult moments we seem to have at work and in life start with our belief system that is obviously immensely impacted by our life experience and our past. Beliefs are thoughts that you keep thinking and these thoughts get translated into emotions because emotions are the "feeling portions" of our thoughts.

If your team is paralyzed by fear and anxiety, it will be extremely difficult to move forward. They will get stuck in their emotions and focus on blaming, moaning and whining, rather than moving into possibility thinking to get the creative juices flowing. **The way many businesses operate today ensures that employees are in a constant state of fear. They feel threatened, undermined and defensive, yet we expect them to think clearly and be productive.** If you choose to ignore this serious issue at your workplace, you can expect people's health and productivity to suffer, as will your bottom line.

Viewing a problem from a different perspective and allowing your team to play a key role in the problem-solving process will not only lift their spirits and increase their motivation, it will also increase their self-esteem and energy level. Because people learn best when they face new challenges, it is important to teach your team responsibility and ownership by encouraging them to find solutions instead of solving the problems for them. If you not only accept but appreciate the creative process of possibility thinking, your team may surprise you with their innovative ideas and solutions.

The era of job security, complacency and resistance to change is over. It is time for every single employee — from the shop floor to the executive suite — to focus on themselves and their contributions to the success of the business. Going forward, we will not only need to encourage accountability, we will have to have the courage to hold each other accountable for our own behaviors, actions and reactions, and have the maturity and humility to be held accountable.

KARIN LINDNER ❧ 51

If we don't believe that we can be the market leader in our industry, we won't be.

If we don't believe that we can keep manufacturing in North America, we won't.

If we don't believe that we have the power to choose which path to follow, we won't.

If we don't think that it is within our control to create a new manufacturing era, it won't be.

At the moment, I see countless companies that are overwhelmed with the waves of change. They are almost paralyzed with uncertainty about how to deal with this new and unknown situation. The question is how can you shift from crisis to opportunity? How can you know, go and show the way as a leader? I would say the same way you would eat a 16-ounce steak: one bite at a time!

Here is my advice to get off to a good start:

- **Take the time to think, scan the horizon for a picture of the future and have the courage to chart a new course.**

- **Invent the future instead of trying to redesign the past.**

- **Embrace the creativity of your workforce.**

- **Listen with an open mind to what your workforce has to say and learn from them and their perspectives.**

- **If you don't have the resources, use the resourcefulness of your whole team.**

- **Be humble and keep learning and growing.**

- **Choose — deliberately — to be positive, optimistic and enthusiastic.**

The sooner you embrace a different approach, the better able you will be to deal with rapidly evolving circumstances. Organizations that refuse to change, or change too slowly, will have even bigger problems. They won't survive in the Age of Instability.

The main reason that excellence is often talked about but not often demonstrated is that the majority of us are rarely challenged or led in this direction. **For most people passing is the goal, excelling is not.** Very few of us have the self-drive and discipline to go beyond average.

I sincerely believe that developing passion for excellence starts with small steps. It means having an attitude that continually seeks to improve. It is about striving to get better as a leader, co-worker, spouse, partner, parent or friend, every single day of your life.

Here are some questions for you to consider:

- **What do you do to get better?**

- **When was the last time you took a critical look at yourself and your own level of excellence?**

- **Are you a role model for other people (without being a legend in your own mind)?**

- **What was the most recent mistake you made or miscommunication you had?**

- **Without blaming someone else, why did it happen? What could you have done to prevent it?**

- **How can you do better next time?**

The direction North America takes will largely depend on our corporate leaders and the vision they are able to create. Keep in mind that in a leadership role, competence goes beyond education, skill and knowledge. It also encompasses behavior, attitude, mindset, and the ability to positively influence others. Leaders who are continuously willing to improve their levels of competence will not only earn the respect of their colleagues and employees, their whole life will be positively impacted. **I always say, "If you want to be competent, you have to have the courage to learn from your own incompetence."**

Today's leaders must have courage to move in the direction they've charted. If ever they look in the rearview mirror, it should only be to make sure that no one is passing them.

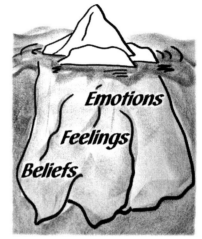

Chapter 7

MAKE DIVERSITY AND UNUSED BRAINPOWER NEW CONCEPTS FOR SUCCESS

"If everyone is thinking alike, then somebody is not thinking."

~ George S. Patton

Have you ever noticed that it is easier to complain about what we don't have instead of appreciating what we do have? Would we rather complain about current circumstances instead of being open-minded enough to see the possibilities that these circumstances can bring about? **What would you think if I told you that by changing your perspective, you would be able to see things that you couldn't see before?**

How about taking a deep breath, slowing down, and standing still for a moment to focus on the good things you have and the opportunities you can create in your current business environment? What a great concept this would be if it worked, right? Well, I promise that it will.

For the majority of people who work in a fast-paced manufacturing environment, it's difficult to imagine slowing down and envisioning better and different ways of doing things to increase productivity. Many organizations extol their employees as their greatest asset and rightly so; employees can either make or break you. If a company has a strong employee base, its strategies will be in sync with the realities of the marketplace, the economy and the competition. One thing is certain: If you don't get the human side of this process right, you will never realize the full potential of your business.

I have adopted the mindset that I can learn something from every person I meet. When people attend my workshops, whether they are management, supervisors, team leaders or other members of the workforce, I always see them as a "10." I believe that they will show a sincere interest in what I have to say and that every single person in the room has the potential to go above and beyond the confines of the classroom. I want to learn from them and they, in turn, learn from me. That's my intention. I see it as my job to increase their energy level and bring out the best in them. I believe that most people want to do a good job, they want to be involved, they want to contribute to society, they want to learn how to be better at what they do, and they want to learn ways of becoming better people in general.

Isn't it mind-boggling that most employees can never reach their full potential because so few managers know how to bring out the best in them?

Not too long ago, I spoke with a young VP, the son of the owner of a manufacturing facility. At one point during our conversation, he said, "Karin, I honestly think that most of my workers are stupid. Some people have 'it' and some people don't. Most of the ideas we get from the workers, which are not many, I find to be stupid ideas. Seriously, how creative can a guy on the assembly line really be? They want to spend minimum time in the plant and they want to get the highest possible pay. That's it."

Three very simple questions sprang to my mind. The first one was, "Why would you hire stupid people in the first place?" The second was, "How do your employees manage to stay in the job if they are stupid?" And finally, "Did they become stupid after they started working for you?"

I then asked him the following: "How many managers and supervisors do you know who make a sincere effort to bring out the best in their employees? I don't think people are stupid. They need guidance, leadership and support on how to become better at what they do and how to become better individuals. I have met more open-minded people on the production floor who willingly adapt to change than I have met in managerial positions. What actions do you take to address this issue?"

He couldn't really answer my questions, but if he thinks that his workers are stupid, I have to wonder what his workers think of him. There is a difference between being unknowledgeable and being stupid. Being unknowledgeable can be corrected through education and training. Stupidity is a lack of intelligence, understanding, reason or wit.

I don't even want to imagine how de-motivated his workforce must be. What is the quality rating of this company? What's its health and safety record? How competitive can it be in this accelerated global market? Can it really be that people like this lead organizations? Does this VP honestly think that his company will continue to grow and prosper if he does not have his employees on board? Bosses who think that they're superior and treat their employees as second-class citizens have a lot to learn about leadership.

I certainly don't think that this executive is a bad person. He has the title but doesn't understand his role and responsibility as a leader and, sadly, he is just one of many.

A very wise and successful man once shared this simple philosophy: "If you have a problem in your business, you have to first acknowledge that you have a problem. If you don't acknowledge that you have a problem, this is the real problem." How true. Most of the time we are not aware of our reactions, actions, and the way we think and speak. How can we change something that we are not cognizant of?

Unfortunately, in many cases there is a huge disconnect between the management team and the workforce. Business schools focus on financial management instead of people management. If MBA graduates end up in manufacturing environments without being fortunate enough to have great mentors, this will only broaden the gap between management and employees. They won't understand each other and they won't know how to connect. Both groups of people have challenges to confront, but there is no shared reality. They say that "ignorance is bliss," but when you don't understand each other's world, reaching the next level of growth or progress will be an even greater obstacle.

A lack of communication, information, mutual dignity and respect may make an employee appear "stupid" when in reality his or her unwillingness to help the company succeed may be a result of not being emotionally connected to it. Why would the workforce want to use its brainpower if its input is not invited or appreciated in the first place? This is discretionary effort after all, and employees won't give it to an employer who is not worthy of it. It is a vicious circle.

I do my best to find something positive in every situation. Have you ever considered that STUPID could also stand for Smart Talented Unique People In Demand? It is just a matter of how we label people and the perspective we have when we do so.

What about you? Do you ever feel exhausted because your employees don't do what you want them to do? If this is the case don't shoot the messenger, but your team is a reflection of you and your leadership capabilities or lack thereof.

It is human nature to like people who are similar to us and have a similar way of thinking, but if we learn to broaden our horizons and to appreciate those who have a different perspective, perhaps really try to understand where the other person is coming from, it can open up a whole new vantage point. That's when real innovation happens.

We cannot continue micromanaging our employees to death and not allowing them to fully use their brains. I believe many people are not using their brains because they are simply not allowed to. A lot of managers still enjoy demonstrating their so-called power, and feed their egos by putting employees down instead of encouraging them to step up with new and innovative ideas. **My common sense tells me that no one knows how to improve the work better than the people who do it every day.**

With our diverse workforce of people from China, India, Eastern Europe and many other places, we have the global advantage right at our doorstep. In my line of work, I regularly meet people of various ages and communication styles from different backgrounds, religions and cultures. I feel lucky to know them. I have not only learned to embrace diversity, I see it as an incredible advantage and opportunity to create a completely new dynamic within our organizations. In the Canadian province of Ontario, we have the distinct advantage of a highly diverse workforce. We have people from virtually every part of the world who have brought their education, experience, and willingness to work for the opportunity to build a better life for themselves and their families. We should be tapping into that eagerness and drive! They've often told me that they have ideas and solutions and know how things could be done more efficiently, but nobody bothers to ask them. Why?

Many were at one time engaged employees but as a result of how they were treated—or perhaps ignored is a better word—they have just tuned out. They now come in, do their job and leave. How many opportunities do you think we allow to slip away by not utilizing the brainpower of our diverse workforce? In my opinion, the problem boils down to how we view or label people, especially people from different backgrounds. We can make really poor assumptions. Assumptions can hurt us.

Here is an extraordinary excerpt and statistic from Dr. Bob Nelson's book **Keeping Up in a Down Economy: What the Best Companies Do to Get Results in Tough Times***:*

> "Do you know that the average American worker makes 1.1 suggestions per year on the job — one of the lowest rates of any industrialized nation? Compare this to the average Japanese worker who makes 167 suggestions each year and you see the potential that exists for great workplace ideas — if you could only find a way to tap into it with every employee."

Have you ever wondered why it is sometimes so difficult to accept other people's ideas, regardless of what culture they come from? It seems particularly challenging if the idea comes from someone who, in our mind, is hierarchically below us. **The pervasive mentality remains, "I am the boss, I am smarter, I know best, and I have to know it all."**

The key here is being open-minded. Consider all perspectives as possibilities. This does not necessarily imply that one agrees with all of the perspectives. Rather, true open-mindedness is a willingness to accept that others' beliefs are as legitimate as our own. Despite our daily demands, it is important to be open and aware of new and different ways of doing things. I know that I am guilty of not always appreciating this myself. Many times I think that I have to do it all on my own, instead of accepting and appreciating other people's help and input. It is difficult for me to accept help, but I am learning. Being open-minded will help you be more flexible and better deal with change because you are more willing to change your perspective. Open-minded people are less judgmental, more optimistic, have great problem-solving skills, and are always willing to learn and to grow.

I am a firm believer that there is a lot of unused brainpower in many organizations. Employees in general, but front-line workers in particular, are still viewed more as bodies than brains and that's why many companies don't reach their full potential.

I am convinced that every worker has a $50,000 idea in his or her head, but there are three main obstacles that need to be overcome:

- We have to find a way to draw out the idea.

- We have to have the intent to implement as many ideas as possible.

- We have to empower employees to implement their ideas, even if it means that they have to find cost-saving ideas in order to make the implementation of their idea possible. For example, many times employees get shut down because they are told that their idea is too expensive. Instead, say, "Even though this won't be cheap, it's an excellent idea. Where do you think we can save some costs so that we can implement your idea?"

If you want to discover and tap into the brainpower of your people, you have to show them respect. This means you make them feel valued, appreciate the diversity they bring to your team, ask great questions, listen to what they have to say, and give them your attention. If you respect your employees through sincere face-to-face conversations and genuinely listen to what they have to say, you will really go places. **Continue to share your wisdom, but remember to empower rather than overpower.**

Chapter 8

WHY MINDSET
MATTERS

**"Understanding the cause of failure is
important. Understanding the cause of
success is far more powerful."**

~ Ed Oakley

How easy would it be for you to define the cause of success? I
guess first we would have to define what success is, and obviously
it means different things to different people. I leave it up to you
to discover your own definition of success.

One of the things that I have noticed in the corporate world
is that everyone seems to be searching for the causes of failure.
I have to ask myself why no one seems to be searching for the
causes of success. I suppose it's because if we evaluate why something didn't work out, it gives us a chance to whine, moan and
complain. On the other hand, in many organizations it would
surprise most people if you asked them to share the causes of
their success with the group.

Focusing on targets, profits, numbers and facts is the norm in
the majority of workplaces and most leaders seem to forget that
in order to get the numbers, they have to have their employees

on board. Based on my experience, there are many people in manufacturing who have very analytical and logical minds, and it almost seems to be part of their DNA to require proof in order to believe that something is possible. Facts are certainly important in any business provided, of course, that facts and data are not the only things that companies focus on. Employees need to be encouraged to take risks and to make things happen. In my opinion, **FACTS** can also stand for **Fear And Complacency Threaten Success.** Sometimes we refuse to believe things we don't understand, but this is also how we create our own limitations. It is so much easier to focus on numbers and results than on mindset, attitude and behavior.

A while ago, a General Manager said, "Karin, we know what has to happen and what needs to change, but what we know and what we do are two very different things. There is simply no sense of urgency and as long as we can drag it out, we will."

I like to call this the "Achilles Factor." As a marathon runner, I experienced serious problems with my Achilles tendon. An Achilles injury is extremely painful and takes time to heal. While I realized that it was definitely bothersome, I certainly did not have to dial 911. It wasn't as if I had broken my leg because if that had been the case, I would have had to go to the hospital. With an Achilles injury, however, I had a choice to make. Did I want to keep running and ignore the pain? Did I want to take a risk and have it get worse? Did I want to keep hoping that it would heal on its own or did I want to be proactive, get the proper treatment from a physiotherapist, and do the recommended daily exercises in order to get better as quickly as possible? I decided to work with a physiotherapist because my injury would not have improved had I not followed his recommended course of action. A physiotherapist can only tell us what to do but we have to actually do it. The best doctor, the best physiotherapist, the best trainer, the best coach cannot help if we fail to heed their advice and follow through with it.

The same holds true for many businesses. As this General Manager mentioned, in many instances, there is no sense of urgency. There seems to be a payoff to stay where we are instead of moving forward. I have to admit that it can be very tempting to do nothing and simply hope that the situation will improve. However, nine times out of ten the problem persists and gets increasingly worse.

So what's the best way to prevent the Achilles Factor in your business?

- **Make sure that your business is in good shape overall by being proactive instead of reactive.**

- **Acknowledge problems as opportunities to learn and to grow.**

- **Educate, educate, educate to make sure people can reach their full potential.**

- **Maintain good morale within your organization by listening to what your employees have to say.**

Don't think that you have to wait to "break a leg" before you can take action!

Consider Henry Ford when he decided that he wanted to build his now famous V-8 engine. He worked against the odds and all the experts told him that it was impossible. Ford's formal education was limited, but his desire, his strong will and his persistence enabled him to succeed. That's exactly the mindset we have to develop and we must teach it to our colleagues and employees.

After years of research, world-renowned Stanford University psychologist Carol Dweck wrote an excellent book, *Mindset — The New Psychology of Success*. She explains that people have either a fixed or a growth mindset. Mindsets are just beliefs that you may have in different areas of your life but the type of mindset that you have overall will play a key role in determining how you respond to failure. People with a growth mindset don't see the bumps along the way as failing; they see them as learning opportunities to work harder and to put more effort into succeeding. The great news is that everyone can learn to have a growth mindset. Simple awareness can already be encouragement enough to get people to think and act differently. Can you imagine the possibilities?

Whenever I facilitate a workshop with a group of shop-floor employees, it never ceases to amaze me what is possible when you show a sincere interest in people and not only listen but hear what they have to say. I show a genuine interest in their challenges and concerns and uncover the real reasons for their frustration. I know I am getting somewhere when the energy in the room starts to shift. What do I mean by that? People enter the room and many of them carry baggage of negativity, frustration and stress with them. Their energy levels are low. They are hesitant about what to expect until I start building trust and peeling off emotional layers to get to the root cause of what's really on their minds. After only a two-hour workshop, management is able to recognize a significant attitude and behavior change within the attendees. They walk, talk and act differently. Is there a secret? Not really. By acknowledging that everyone wants to get the best out of his or her own life and by teaching employees how to deal more effectively with certain situations and listening to their concerns, I am able to hit a home run with the majority of people. **My goal is to teach them to come up with their own solutions instead of having them wait for someone else to solve problems for them.** Most of them are not used to that. Do you think the same applies on your shop floor? That's exactly what you want; you want the energy to shift in order to develop a success-oriented mindset within yourself and your team. In order to make this transformation sustainable, I have to rely on management's trust and open-mindedness to support the process and apply my simple hands-on tools and techniques.

- **Why don't we think that the pursuit of perfection is everyone's job?**

- **Why have we settled for mediocrity?**

- **Why are we trying to secure the right answers instead of pursuing the right questions?**

- **How do you think your bottom line is impacted when your employees are negative, frustrated and stressed?**

If you want to succeed, do not label people and do not give them the feeling that they are inferior and offer no value just because they work on the factory floor. They are the root cause of any success because production is where the money is made. These are the people who have to ensure that a quality product leaves the plant, these are the people who give 100 percent and more if you don't take away their self-worth, and these are the people who are the specialists in what they do. They can offer incredible assistance and support in the area of continuous improvement and innovation. Isn't it our responsibility to help them to look at their jobs with different eyes? How do you view your workforce? If you want to explore the cause of success, go to your employees, talk to them, build trust, and listen to what they have to say. Making them part of your success is the secret to your success.

Employees tell me all the time that they only see their managers when something goes wrong. Leadership authority Ken Blanchard has always pointed out that we have to catch people doing something right. Why is this so difficult to put into practice?

Follow successful organizations as role models and simply do what works for them. It sounds easy but the best practices of other organizations are not enough. You have to create the culture that supports those practices, and that takes energy, effort, self-discipline and determination. To make it work, you may have to start with yourself and your own leadership skills. It takes commitment, patience and persistence. Where's the fun in that? But what's the alternative? Would you rather stick with what you have and be stuck in the status quo?

Truth be told, it is quite scary when it comes to working on our own mindset, habits and behaviors. Many people fear this kind of change. However, you shouldn't look at it as being one overwhelming task, but rather as a process of making small improvements one step at a time, one day at a time, so that you are able to lead by example and become the best leader you can be. In my opinion, it is actually highly rewarding and a lot of fun once you can reap the first fruits of your labor. Remember, your team can only be as good as you are!

Due to daily operational pressures, we have too many disconnected leaders who are far removed from the problems and challenges that their managers have to deal with, let alone the issues of the shop floor employees. Perhaps it doesn't even cross the minds of most CEOs just how significantly this disconnect affects their overall success in business.

We have to shift people's mindset from "I can't," "It won't work," or, "That's too expensive," to a mindset of "How can I?" "How can we make it work?" "How can we find the resources to make it happen?" or, "How can we tap into our collective resourcefulness?"

Henry Ford's mass production system was a revolution during his time, but times have changed. Each job that can be replaced by a robot should be a wake-up call to everyone about the necessity of continuous learning, self-development and adaptability. We need to encourage people to fully engage their brainpower. Shouldn't we be excited to develop a culture of curiosity and people who look for opportunities and possibilities throughout the day?

It is my firm belief that you have to incorporate your thoughts into your business strategy. Our thoughts have immense power. They affect our personal and our professional lives and we should use this knowledge and take advantage of it. It may be too much to start monitoring our thoughts, but let me tell you from my own experience, it can help you immensely if you just create more awareness around your thought processes. Successful people don't take no for an answer. A no for them is simply another obstacle to overcome.

The way you do anything is the way you do everything. How different would your day be if you started it off on a positive note? Just imagine going into your office, work area or workstation thinking that you want to do something nice for someone else. Fire off an email to your team with an inspirational message, talk to a co-worker and give a sincere compliment, or demonstrate an act of kindness by serving coffee to your team or your co-workers. Leave your office and walk around with open eyes and make it your goal to connect with the people. Look for the things you can do to make someone else's day and let people know when they do something well. You may not be able to imagine the impact right now, but these kinds of actions will start to spread like wildfire.

Here are some simple things you can do:

- **Be aware and be "present" in each moment. Take the time to talk to the people around you. It keeps you connected and informed.**

- **Force yourself to see the opportunity in any problem you may encounter. That's the mindset that will help you and your team to not only succeed, but win.**

- **Replace the word problem with challenge. Doesn't that already feel much better?**

- **Encourage people to share their knowledge, wisdom and ideas.**

Be determined to have the mindset to win, to succeed and to start serving your team. After all, your team is the secret to your success. This will help your business stand out and put you miles ahead of your competition. What are you waiting for?

Chapter 9

A PROACTIVE APPROACH AGAINST LOW-COST COUNTRIES

"Some people want it to happen, some people wish it could happen, others make it happen."

~ Michael Jordan

In which category would you place yourself?

If you want to make it happen, you have to want it. Making things happen takes time, energy, commitment and self-discipline. If you want change, it's either because you are forward-thinking and open-minded, or you are in so much pain that there is no other option. Either way, it can be a sustainable road to success!

One of the greatest challenges is to change beliefs regarding what works and what doesn't.

When I asked industry leaders how we can take a proactive approach against outsourcing to low-cost countries, the majority of the responses I received led me to believe that most people seem to have accepted the status quo. They feel that's the road they have to follow in today's business environment. That's just how business is done now. If we don't like certain circumstances, do we have to accept them? I don't think so, do you?

Think about this example: There are countless franchise restaurants throughout North America, and fewer and fewer restaurants offer first-class service and uniqueness. Teenagers are growing up in a world where they are learning not to take chances or experience new things. At franchise restaurants, they know exactly what they're going to get before they even enter. Our children are used to a world where individuality seems to have less value. Why bother to try something new? Doesn't this demonstrate how low we have allowed our expectations to go? It's not the fault of our youth. It's our fault as a society through the examples we set. We have created a world with lower standards and most of us accept this because we never take the time to think about it.

The question is, are we comfortable with our own mindsets, actions and the examples that we set for others?

People look for the cheapest price and through the Walmarts of the world, it is achievable. But we need to understand how our own actions affect the economy. **Current global conditions challenge our fundamental beliefs and values. As a result, we experience attitudes and behaviors that lead to unsatisfying performance individually and as organizations and countries. Something, somewhere along the line, has gone very, very wrong.** Unfortunately, we generally have a difficult time acknowledging that we are wrong. **We blame it on our countries, our governments, our companies, our customers, our suppliers and our teams. We take the misguided and selfish view that "I am right and you are wrong."**

To be honest, just a couple of years ago I never thought that my actions would impact anything in the grand scale of things. However, when I learned to think differently, I realized that the only way to start creating a different environment was by taking a critical look at myself and taking responsibility for my actions.

Now I truly enjoy making conscious buying decisions. For example, I think it is important to know where my food comes from and that's why I buy many of my products fresh from local farmers or at the farmer's market. My upbringing in Austria certainly helped me develop this kind of attitude because my parents were role models when it came to buying local produce, dairy and meat. That's all we ate, either from our own garden or from local farms. My mom repeatedly said, "You are what you eat." I prefer to pay a little bit more rather than eat something that comes from China. I feel that if I do my part to support local farms, it will help to keep those farms in my community, ensuring we can continue to have access to nourishing, tasty and abundant food. For me, it is also about raising the consciousness of people in my immediate environment. Over time, it has become my habit to check every potential purchase to see where it is made, and I have come to realize that my actions are contagious. Many of my friends now do the same. To be honest, sometimes it is a real challenge to find certain products that are made in North America. It takes time and effort to uncover alternatives for imported goods, and it certainly takes the willingness to pay a little bit extra.

While our American neighbors have created the brand "Made in America," Canada doesn't do a very good job promoting home-made products. What I find most disturbing is that most of us don't even know what "Made in Canada" means. Is it manufactured here or only assembled here?

I am a huge Apple fan, but since reading about suicide nets and the terrible circumstances of factory workers in China, I have shied away from buying an iPhone. What stops Apple and many other companies from manufacturing here? Lower profits, of course! Many people are vegetarians because of the way humans treat animals. I wonder how many people would give up their fun gadgets knowing about the inhumane way we treat other human beings, just because they are out of sight in another country. Probably not too many. The main thing is that we do well here, right?

In the German language we have a word for that: "Scheinmoral," which means hypocritical moral.

A little over three years ago, I wanted to buy a North American car. My simple requirements were safety, quality, great design and fuel efficiency. None of the original Big Three could deliver what I was looking for. As much as I am ready to spend money for a product that is made in North America, if the quality is not there, my loyalty seems to disappear. My next car purchase will come soon and I will make sure that whoever wins me as a customer also provides jobs for people in Ontario. Those are just a couple of simple actions that I set in my personal life. Some care, some don't. Do you?

During the interview process for writing this book, one of the things I asked was, "In what ways could we start differentiating ourselves from countries such as India and China?" Every single one of my interviewees leaned back in his or her chair and I watched their eyes drift toward the ceiling. As they pondered my question, they came up with answers such as innovation, continuous improvement, more and better "processing" techniques, quality, customer service, face-to-face contact, larger and more complex components, development of new processes, specific project management systems, and last but not least, a focus on high-end manufacturing.

As you can see, it is possible to beat low-cost countries on so many other levels, but if we only focus on the cost, we will bite the dust in the long run. It is impossible to compete against low wages. The hourly cost of a North American worker is ten times higher than the hourly cost of a Chinese worker, at least for now. Still, many manufacturing companies in North America don't feel any social responsibility or commitment to their communities. **The workers feel cheated out of their livelihoods when manufacturing companies move away and yet these very same companies expect these displaced workers to buy the goods they're having manufactured cheaper somewhere else.** Many of these workers are now either out of a job or work in a job that pays half of what they used to earn.

On the other hand, we cannot forget that most people, including factory workers, were able to significantly increase their material well-being by purchasing consumer goods from China, Indonesia, and so on. That's why awareness and taking on personal responsibility for our actions is key. **I want to make it very clear that taking responsibility for our future has to be shared by companies and consumers. How are your actions in not just one but both camps influencing our future?**

Doesn't it seem crazy to move our jobs and knowledge, and with that our future, to low-cost countries? **How can we ever expect our economy to recover if we lose our middle class?** Let's start thinking in the paradigm of our own family. Is this the future we want to create for our children? Shouldn't we do everything it takes to keep our knowledge, wisdom and, most of all, our passion within our borders? The countries that are doing well and that have survived the economic downturn all manufacture something. Do we really want to risk losing our wealth and prosperity by slowly developing into a service industry? The shocking reality is that many people seem to think that we will be okay as a service-industry economy, but we won't. Where would the income to pay for these services come from?

In short, we must use our intellectual capital to sell high-quality solutions to customers because that's what they want and expect. They are looking for the most cost-effective solution to problems. Innovation and creativity are certainly on top of the list if we want to stay ahead of the competition. Everybody knows that. Everybody says that. In the end, though, everybody is waiting for somebody else to take action and very few people are actually doing something about it. It is far from easy to take on some responsibility and take a step in the right direction. **Our mindset has to change 100 percent. We must create an environment with unique products and there is no doubt in my mind that this can be done.**

For example, companies in China and Vietnam are fabulous at replicating and improving, but they are not as good at innovating and creating. This is largely because they were not permitted to do so during their experience under communist regimes. Many people I spoke with seem to think that China would rather leave the innovation process to another culture, but I believe that they will, in fact, focus and channel their energy on becoming more innovative and creative. That's an ability that has always fascinated me about the Chinese culture: their eagerness to learn and their willingness to take immediate action. It's certainly a learning process for all of us.

In my many conversations about this topic, one thing became very apparent to me: Few leaders have the desire to create change for the better because it comes with the expense of time, energy and effort. Where is the hunger for success that is the key for long-term sustainability? The responses from my interviewees proved that it is theoretically feasible to stay ahead of the competition no matter where in the world the competition is located. It comes down to having a competitive spirit and wanting to stay ahead of the game.

I heard one comment with respect to quality that really made me think. This individual said that there is no doubt in his mind that the Chinese will soon catch up with us in terms of quality. We may only have a two-year window. Wouldn't it be better to stay ahead instead of accepting what many may think is inevitable?

Allow me to again compare running a business to my experience running marathons. My strategy is to stay ahead of the people who are slower than I am. I don't want them to catch up. They may eventually get better, yes, but so will I. Shouldn't this also be true in business?

Are we working ourselves out of our own jobs? Low-cost countries are driven to gain knowledge and experience provided to them all too often by their North American customers. They're eager to learn, grow and get better. In response, we have to have to have confidence and get going with a dynamic "can-do" attitude. If we focus on improving ourselves, as we've discussed earlier, leading through example, inspiring others and changing our mindsets, I have no doubt that we "can do" it!

Chapter 10

THE POWER OF CULTIVATING WORKFORCE AND LEADERSHIP BEHAVIORS

"If we did all the things we were capable of doing, we would literally astound ourselves."

~ Thomas Edison

I am disappointed when I see people working in a poisoned environment, toiling in negativity day in and day out. It not only has a detrimental impact on their productivity, but more importantly, on their health and well-being, all for the illusion of security and a regular paycheck. Why do so many organizations ignore their crippled company culture and choose to stick with the status quo?

It's time to wake up! We need to realize that we have to compete in order to survive. **We have to keep getting better plant by plant, leader by leader, and employee by employee.**

Working with companies in unionized environments has taught me some valuable lessons. However, there have been occasions when some of the comments that I have heard from employees have left me at a loss for words. One young man told me that he thinks that a factory worker who has worked for a company for more than 25 years has earned the right to sleep on the job. I am sure he has never considered the consequences of sleeping on the job or the impact his thinking has on others and the company he works for. Is this what the world has come to? I did my best to explain to the employees in this workshop that if North American companies want to be known for their quality products, innovation and creativity, and if they want to compete globally, they simply cannot afford to have employees who sleep on the job. Let's not even talk about the health and safety issues this presents.

Furthermore, there are companies that allow employees to get away with unacceptable behaviors because supervisors and managers do not know how to effectively address these situations. It is much easier to hold people accountable for their results than it is for their attitudes and behaviors. The fellow I just mentioned may not be bothered by a slumbering co-worker, but in general this would frustrate most hard-working, dedicated employees. While I am not keen on having too many rules and regulations, I know that in their absence, groups will make up their own.

In another workshop, an employee came in, sat down and promptly decided to have a snooze. His company is not unionized, but it does have many employees with high seniority. It is important to recognize that employees who are disinterested and who have tuned out will try whatever they can if they think they can get away with it. In my case, he saw a blond woman at the front of the classroom and thought I was an easy target. Unfortunately for him, he didn't know me or how straightforward I can be. I addressed the problem immediately by asking him if he was being paid to attend the workshop. When he said "yes," I told him that he had a choice: to either sit up and participate or go back to work. The room fell silent. His co-workers and fellow participants glanced back and forth between him and me. He decided that he would rather participate and he did. As a result, I not only earned his respect, but also the respect of everyone else in the room. After the workshop, he approached me and apologized for his initial disrespect and told me that he had really enjoyed the entire class. This is one of the small wins I like to celebrate. He approached me because he wanted to, not because I forced him to. He realized that his behavior was wrong and would not be tolerated.

If you compare gardening to managing people, there really is no difference. The seeds you plant now you will harvest later. People demonstrate negative attitudes and behaviors because they're accepted. Our leaders are swamped with so many demands that it becomes a real challenge to not only address these issues but also to take the time, energy and effort to create sustainable change. There is such a disconnect between many corporate leaders and their workforce that they don't even realize that there is one. If they do, they don't know what to do about it.

If each employee was committed to making small, sustainable improvements, tremendous operational results could be achieved. Very often it comes down to personal awareness in order to overcome a certain roadblock. Most companies are so busy measuring productivity that they can't see the real reason or root cause for the loss of productivity within their organizations.

A good analogy is energy efficiency. If you heat your house, you want to ensure that it is energy efficient. If the windows and doors are not properly sealed, the house will never be as warm as it could be and your heating bill will be ridiculously high. It's the same with managing people's energy. If you are a manager, you want to make sure that you understand what your employees do, how they do it, and how they can work better by reaching their full potential. This will be time and money well spent.

Why do so many managers think that employees will be more productive if they are controlled?

Well, sadly, there are a lot of people who will work a little bit harder when they feel they are under the influence or control of someone else. However, you cannot possibly control people all day long and if you try, you will be frustrated, exhausted, and won't see the expected results in the long run. There will always be opportunities for employees to slack off and do other things if their jobs don't have any real meaning or purpose for them and if they don't feel any connection to you or the organization.

So why do I think it is important to give people their independence on the job? Won't they take advantage of our trust? First of all, I really don't believe that they will. Secondly, I don't know many managers who have regular conversations with their employees about honesty, integrity, expectations, and the importance of having their 100-percent support. **Most people don't know what their managers' expectations are, and if they have the perception that management doesn't care, they will only do the bare minimum.**

On the other hand, management seems to wrongly assume that just because employees are getting paid, they will automatically care enough to give their all for their company's success. Sorry to be the bearer of bad news, but that's not how it works. If you want to cultivate your team, you have to start changing your approach.

Managers have to first establish high standards for themselves and be role models for their employees. This may push some individuals out of their comfort zones, but that's a good thing! **Keep in mind that employees are observing managers' actions and reactions on a daily basis. To be a leader you have to earn the respect of your employees, and this is only possible if you lead by example.**

The global marketplace is shrinking and the war for talent is raging. Recruiting talent is one challenge, but retaining talent is even more difficult. I think that it is time to discard what we think we know and start to learn what we have to learn. Where are the common-sense human behaviors such as respect, care and compassion? Let's focus on learning how to identify and understand certain emotions in a conversation and make the other person feel valued and understood. How many people are really capable of doing that? If we are not able to lead by example and empathy, how can we expect people to behave differently?

I find that basic life skills are missing in our corporate world. **Everything that we work so hard to teach our children seems to fly out the window in many workplaces.** Treating one another with dignity and respect doesn't seem to be part of the expectations. We start developing life skills at home. Our environment and the people who have an influence in our life all take part in this development. However, I espouse taking it a step further. More courses on how a positive attitude and respectful behaviors can impact our success in life have to become part of the curriculum in our schools, colleges, business schools, universities and workplaces. Why would we think that teaching people about the power of their own attitude is not important?

Isn't North America the land of endless possibilities? What has happened to the drive, energy and ambition that people used to have? In my experience, it should not matter how often you fall down, it should not matter how often you hear "no," it should not matter how often you try something that does not work. It's about dusting yourself off, readjusting, getting back on track and trying again. We all know that when toddlers learn to walk, they try, they fall, and they try again. Why? Because failure is not an option and they are curious to see the world from a different perspective. They will walk at 9 months or at 15 months, but unless they have physical challenges, it is a given that they will walk. That's the energy and spirit we have to rediscover.

In many European countries, people cycle or walk to work. While this may not always be possible here, let's just imagine for a moment what a regular exercise routine would do for you and your team in the morning. The energy level would undoubtedly go through the roof. Isn't this a better way for people to start their mornings rather than being inundated by emails and negative comments? **Here is a newsflash: Life cannot be just about work!** It is essential to encourage people to look after themselves — from feeding their minds and spirits to proper nutritional intake for optimum health. **If you feel good about yourself, it will be easier for you to deal with all of the challenges you may face throughout your day.**

Oftentimes people are denied the opportunity to excel in their jobs. They aren't challenged to be the best they can be. That's when some choose to start participating in various sports activities; that makes them feel good because they want to see results. That's how it was with me. **As a marathon runner and triathlete, I know that through sports I have discovered many of my own qualities that I would otherwise not have been able to display as an employee. What a terrible loss that would have been for the company!**

Personal interests tell a lot about a person, but in order to discover that, you have to care about the people who cross your path. For example, a marathon runner displays many qualities that could be very useful on the job. They are goal-oriented, have a keen desire to improve, and they possess self-esteem, passion, energy, intrinsic motivation, dedication, persistence and commitment. Wouldn't this be a person you'd like to have as an employee? I am not saying to only look for sport enthusiasts, but every person has certain interests and these interests are manifested in the qualities this person has. It is then our responsibility as leaders to help them improve in the areas where they already naturally excel.

I am convinced that companies who encourage and support their employees to physically and mentally look after themselves will always have a happier and more productive staff. Examples include subsidized cafeterias with healthy food choices; special offers for gym memberships or even an accessible gym at work with classes that can vary from yoga, spinning, karate to salsa dancing; and team sport activities. I have even heard of companies that offer an extra vacation day for people who pass a fitness test. Researchers have found that only one in 13 workers swaps his or her business suit for jogging gear and pounds the pavement for more than 40 minutes while not-so health-conscious colleagues stay at their desks to eat their packed lunches. Who do you think will be more productive over the course of the day? Please be assured that those who venture outside work return to their desks feeling re-invigorated and ready to press on with their workload. Clearly, exercise is good for our health and good for the business.

You may have more than one person like this working for you right now, people who have the desire to add value to your organization. Start looking at your team from a different perspective and help them awaken their spirit.

Chapter 11

WHAT SHOULD MANUFACTURERS RECONSIDER?

"One machine can do the work of fifty ordinary men. No machine can do the work of one extraordinary man."

~ Elbert Hubbard

Although lean manufacturing is acknowledged more readily today than it was just a few years ago, it seems to me that the focus is still too much on the tools rather than the culture. Nevertheless, the direction is the right one.

Going forward, it will continue to be important to streamline processes and reduce costs in order to work most efficiently. Continuous improvement is a must if we want to succeed in our endeavors to compete in this breakneck global market. The only ways to separate ourselves from the competition will be creativity and innovation and they will require encouragement from the top. Standardized processes can hold people back from asking themselves if there is a better way. While standards are important, we shouldn't forget that they require regular review and improvement.

The ability to preserve what worked well in the past and move forward with successful new developments will depend on how well intergenerational groups can work together. The key is the right combination of an appreciation of wisdom and experience, and the acknowledgment of fresh ideas and vision for the future. Mutual respect and open-mindedness at all levels of the organization will also be critical. As long as we honor this, we will be able to overcome roadblocks more easily.

In addition, personal development will be an inevitable part of this path of creativity and innovation. Our manufacturing environments will change drastically as highly automated processes will produce a burgeoning demand for people who are adept with this kind of technology.

You can follow countless discussions and articles on the tremendous need for knowledge workers in our country. Do we really want to continue the pattern of depending almost solely upon the import of knowledgeable and skilled immigrants? Our current workforce must be made aware that it has a part to play in becoming more educated and skilled to prepare for a shift in manufacturing that is already underway. It is our responsibility to continue educating people about the consequences to our marketplace if we miss out on this enormous opportunity. I have seen people at work in manufacturing facilities operating on autopilot. They come in for a paycheck and do their best to survive the day. I am not making a distinction between a manager and a worker; this can be true for both if there is a lack of incentive and involvement.

Why have we developed into a society of entitlement and of being spoon-fed? Who is holding the spoon?

For those who are discouraged and lack optimism, hope and faith are critical if we are to continue on a quest for accelerated achievements in this industry. Employees see that repetitive jobs are being replaced by robots. Knowledge, skill, mindset and the adoption of new habits will become essential for the workforce.

If people are not continuously upgrading their skills and abilities, they may no longer find a job that is even remotely close to the pay range they were used to.

Employees must take responsibility and do their part to become better trained but for them to do so, the right support and systems must be in place. I have spoken to many employees who feel their company owes them increased educational opportunities *and* more money. There is a serious disconnect here. Why not adopt a shared-responsibility model? For example, companies that must ensure that their employees have the opportunity for continuous education, such as apprenticeships or acknowledged accreditations, could share the cost burden with employees. In that way, employees are then also held accountable for their education. This is what adopting a mindset of shared responsibility means. **I would argue that creating a more cultivated, educated and solid skills-based workforce in this way is much more likely to succeed.**

Another point of revitalized interest should be Mechanics' Institutes. Thousands are still operating throughout the world, as libraries, parts of universities or adult education facilities, and in churches, theatres, cinemas, museums, recreational facilities or community halls. However, before my research I didn't even know about them and I haven't heard other people in the industry talk about them. Historically, Mechanics' Institutes were educational establishments formed to provide adult education, particularly in technical subjects for working men. They were often funded by local industrialists and philanthropists on the basis that companies would ultimately benefit from having more knowledgeable and skilled employees. The Mechanics' Institutes were used as "libraries" for the adult working class, and provided them with an alternative pastime to gambling and drinking in pubs. The world's first Mechanics' Institute was established in Edinburgh, Scotland, in October 1821 as the School of Arts of Edinburgh (later Heriot-Watt University), to provide technical education for working people and professionals. Its purpose was to "address societal needs by incorporating fundamental scientific thinking and research into engineering solutions." This school revolutionized access to education in science and technology for ordinary people. Wouldn't Mechanics' Institutes be a great educational forum that could be expanded on for any subject of interest, including inspirational talks, and soft and social skills for management and the workforce? We just have to be more creative.

Continuous improvement can be applied to various areas of our daily work life such as processes, standards and quality improvements, cost and time savings, and product development and design. However, the same level of emphasis should be placed on personal development, especially with respect to expected attitudes, mindset and behaviors for management and the workforce.

The value that manufacturing brings to our society and how it influences our wealth and prosperity is an important message that we need to disseminate to the general public. Adjusting to change and developing into a learning society is what's best for our continent and the future of our children. I am certain that once this awareness increases, people will feel more responsible for doing their share and this will be reflected in their buying decisions. **Now more than ever, we must put immense value on innovation, quality and customer service because this is what will separate us from the competition in the long run.** If we keep investing in North America, we are investing in the future of our children.

During my research, I was fortunate enough to connect with passionate people throughout the industry who sincerely want to make a difference. As my opening quote suggests, there are quite a few extraordinary men and women out there who invest their personal time and money in increasing awareness. Their whole intent is to lead us in a different direction, find a better way, and create a more promising tomorrow.

Creative and innovative business owners, educators, retired CEOs and other movers and shakers in this industry have shown tremendous passion for creating something bigger and better. This pool of leaders will grow and once we start a movement, more people will follow.

Someone I discovered along the way is Harry Moser, a true thought leader when it comes to making the case for "bringing manufacturing back." Moser has been an advocate for American manufacturing for many, many years. His father was his role model. Moser Sr. worked for the iconic Singer Sewing Machine factory in Elizabeth, New Jersey, and it was assumed that his son would enter the world of manufacturing, too. Fortunately, Harry did and now that he has eased into retirement from Chairman Emeritus of GF AgieCharmilles, he has more time to increase awareness regarding "reshoring" manufacturing to our part of the world. It is near and dear to his heart.

"Reshoring is bringing back work, parts or tools that will finally be used in North America," Moser explains. "In other words, we're not saying that you should make everything here and ship it to China to assemble. We're saying if you have an end component that is sold into the North American market or assembled into a product at a North American factory, or a tool that's used in North America, and you're now having that work done overseas, to evaluate the total cost of that subassembly or tool in the States versus overseas. We believe you will decide more should be sourced here."

I too believe in the economic argument for reshoring manufacturing jobs back to North America. The trend has been to ship them out and I know we won't get all of these jobs back, but I trust that we will regain some. Reshoring can't be justified by companies as "just wanting to do the right thing." Instead, there is a strong business case to be made in its favor.

A closer look suggests that off-shored production may mean a cheaper price but not necessarily lower Total Cost of Ownership

(TCO). This means the product's price plus any costs that are jointly incurred by the supplier and the buyer as well as internal costs incurred by the buyer. Free TCO software is available from Moser's Reshoring Initiative at: www.reshorenow.org.

In a case study comparing costs in the United States and China, David Meeker, a lecturer at M.I.T., and his colleague Jay Mortenson found that it is 8 percent cheaper to produce a current design in China. They said, "There are substantial savings associated with purchased parts from China that include direct labor (79 percent savings versus U.S. labor rates), indirect labor and salaries (61 percent savings), benefits (75 percent savings), overhead (40 percent savings) and selling, general and administrative (SG&A) (11 percent savings). When adding logistics to the China price, the cost advantage of producing in China shrinks to 8 percent: $13.85 for a case-study product made in China versus $14.99 in the United States. But when design for manufacturing and assembly (DFMA) software is applied to the same product, the China advantage vanishes. The China cost declines to $9.79 versus the U.S.-made product at $9.47."

Another reason to use TCO analysis is to release the constraints on using methods such as DFMA, TOC, lean, automation and innovation. If a company believes the cost gap is 30 or 40 percent, it is unlikely to try to become more efficient here. When executives understand that the TCO gap is only 10 percent and closing rapidly, the chances of starting and successfully completing such efforts is much higher.

Companies who went full speed ahead and outsourced everything to low-cost countries now need to consider this total cost of ownership. Someone started a trend and too many companies seemed to follow without weighing opportunity or risk. People may have been blindsided by low labor costs and lost sight of all the other risks. Few companies have the courage to admit a mistake in public, but if we learn from our mistakes, isn't that an admirable thing? As long as we don't make the same mistakes over and over again, mistakes can be very valuable. As you know, I believe the more we allow ourselves and others to make mistakes, the more opportunities we will have to learn and to grow. No one should feel diminished by making the wrong choice, as long as we can defend why we made the decision at the time. It's important to be decisive, learn, and take ownership of the outcome without blaming others.

Although I am not a specialist in reshoring, on-shoring or offshoring, logic tells me that it makes sense to produce where you sell. It gives us more control over what we are doing. There are other risks in manufacturing elsewhere as well: currency fluctuations, natural disasters, changing customs regulations, fines and penalties, logistics, poor infrastructures that lead to delays and quality issues, political and social instability ... The list is lengthy, yet often seems to be ignored.

I recently spoke with a project engineer who is currently dealing with a project in Asia. The language barrier with his counterparts in China adds to his frustrations throughout the day and he believes that ultimately, more money is being spent than saved on this overseas project due to quality issues. If we look at all the costs and total them up, chances are pretty good that we can manufacture here just as competitively and with a lot less risk. What's important is the design work we can deliver, our focus on innovation, and implementing more streamlined and automated processes.

Organizations are often too focused on what's going on internally instead of stepping outside "the box" to see what's going on in the real world and what they can learn from others who have already been there. Taking this approach could help them avoid having to reinvent the wheel.

There are many new, exciting and innovative ideas springing from small companies that will tremendously benefit our economy, but isn't this what North America is all about? Ideas created out of entrepreneurialism, passion and the drive to make them happen.

Your job is to be open to new ideas, take them in, and develop them further. Or take it one step beyond: be creative, encourage creativity, and bring out more innovative ideas. By networking with dynamic thought leaders, most organizations can take the quantum leap necessary to becoming better and better every day. There is no reason why you cannot take advantage of the framework that's already out there waiting for you.

Chapter 12

WHY WOMEN SHOULD PLAY A VALUABLE ROLE IN MANUFACTURING

"Lay down your sword and your battle gear, this is not a war. It's more like a blind date with both of you in blindfolds. The ultimate goal is to build partnerships, show your smarts, not your insecurities."

~ Susan Gyopar

The manufacturing sector is facing a tremendous shortage of skilled employees. As a result, it can no longer afford to ignore the desire, talent and energy of anyone, male or female, who is passionate about making a difference in this vital industry.

The best companies I've worked with value diversity. Great results always come when a team of men and women of different ages, races, cultures and education work together. A diverse team collaborating on a project promotes creativity, turns talkers into listeners, and helps everyone look at things from different perspectives. These companies recognize that it is not only beneficial but also highly advantageous for their organizations and for the economy to promote women to senior executive roles.

I believe it is critical for leadership teams, including those in Human Resources, to understand why employing women is good for their business. Money matters and most companies won't necessarily do something just because it is the right thing to do. They have to see the facts when it comes to hiring and promoting women to a corner office. The same holds true with regards to Operations, Quality, Purchasing, Sales and Marketing, and every other department. Performance standards, operational and personnel policies must be developed to create a healthy, fair, and consistent work environment for ALL. When developing roles with women in mind, all employees will benefit and companies need to see this as an overall win that will increase engagement, productivity, efficiency and quality, and improve customer service.

Women in the manufacturing industry are getting better at weathering market cycles and overcoming roadblocks. They are gaining greater confidence in showcasing their leadership abilities. Manufacturing has been slow to move women up in the ranks, but more and more women are eager and driven to take on executive positions in this industry. Over the past few years, the percentage of women in these male-dominated fields has significantly increased, although there is definitely room for improvement. There are not enough women who consciously seek this career choice and it is important to explore why. It is an ongoing challenge to attract women to manufacturing. We have to inspire diverse individuals to enter the manufacturing workforce and eliminate stereotypes in order to improve the overall reputation of this industry.

Mentorship and sponsorship programs for women as well as men in technical environments would be an enormous benefit. Mentorship programs could be established for young achievers with high potential; employees who demonstrate initiative, intelligence and capabilities that set them apart. These are the people who should be nurtured for promotion, greater responsibilities and succession planning.

If I had a nickel for every time I've heard someone say: "Don't get me wrong...I love women," I'd be rich. These are sarcastic words spoken by those who, consciously or unconsciously, want to make sure that women are kept in their place. The possibilities, fears, hopes and consequences of women attaining more leadership positions can be debated to the ends of the earth. The facts speak for themselves: **Women are still grossly underpaid, in spite of having higher levels of education and proven performance records. Competence isn't tied to gender, is it?** According to the following statistics, the number of women in top positions in business is still ludicrously low — certainly as a percentage of competent women in the workplace.

Women are under-represented in many segments of the manufacturing industry. Deloitte Development and The Manufacturing Institute discovered that gender and diversity initiatives are areas of focus in less than 25 percent of manufacturers surveyed. According to recent Catalyst research, the percentage of female CEOs in U.S. businesses is 3 percent; females on American Boards of Directors represent 15.7 percent; female Executive Officers weigh in at 14.4 percent; and the U.S. labor force as a whole is comprised of 46.7 percent women.

If you think these percentages are unbelievably low, particularly given that almost half of the American work force is female, you haven't yet seen the statistics for the U.S. manufacturing industry.

In U.S. Manufacturing — Durable Goods, **women represent 1.1 percent of CEOs**, 13.7 percent of the Boards of Directors, 10.4 percent of Executive Officers, and the manufacturing industry labor force is comprised of only 24.4 percent women.

According to Deloitte, below 25 percent of surveyed women in automotive agree that their organization demonstrates an above-average effort to retain women. Fifty percent believe that performance standards are inconsistent between men and women; of these, 90 percent believe performance standards are higher for women. Nearly 75 percent are neutral to negative on how accepting the industry is of family commitments.

Little progress has been made over the last five years for women in automotive. Less than 25 percent are aware of an active recruitment program for women in their companies. Overall, less than 20 percent of survey respondents rate recruiting efforts as being better than average. Similarly, 85 percent rate female development efforts as average to poor.

Fortune 500 companies with historically higher percentages of women officers experienced 35 percent higher return on equity and 34 percent higher total return to shareholders than did those with low percentages of female corporate officers.

According to a 2011 study conducted by the University Wisconsin-Milwaukee, **only 11 percent of practicing engineers are women.** In a survey of 3,700 women who had graduated with an engineering degree, workplace climate was cited as a strong factor in both the decisions to leave and to stay in engineering jobs ("Stemming the Tide: Why Women Leave Engineering," Nadya A. Fouad, Ph.D. and Romila Singh, Ph.D., University Wisconsin-Milwaukee).

Nearly half of survey respondents said they left because of working conditions, too much travel, lack of advancement opportunities or low salary. Those who left were no different from current engineers in their interests, confidence in their abilities, or the positive outcomes they expected from performing engineering-related tasks.

The decision by women to stay in engineering was influenced by key supportive people in the organization, such as supervisors and co-workers. Currently, women engineers in companies that value and recognize them for their contributions and who invest substantially in their training and professional development expressed the highest levels of satisfaction with their careers.

Globally, women remain the single largest group of humans to be abused and discriminated against. This must change and change is happening, but at a snail's pace. For example, behaviors such as racial, sexual or verbal harassment must not be tolerated. Appropriate supervisory and leadership role modeling must come first. It must be acknowledged that this is a source of stress and intimidation that is still rampant and results in a significant deterrent to women in the manufacturing industry. Unfortunately, there are still too many cases of men in senior-executive positions dressed in Armani suits who may look like a million bucks driving their expensive sports cars but who do not respect people, who resort to vulgar, crude and demeaning speech with their employees and who, because of the positions they hold, feel they have the right to treat others dismissively or badly. Don't you agree that you can't put a price tag on class and character?

There should be a good balance of masculine and feminine traits in the manufacturing industry. I would like to believe that women who choose manufacturing as a career will use their leadership qualities for purpose and passion rather than power and position. I am certain that open-minded and forward-thinking men are ready to embrace more balanced meetings around the boardroom table because men of quality are not afraid of women of equality. Everyone else should fasten his seatbelt because we are entering a different era of leadership.

It is surprising to me that many men seem to think that women don't want to work in the manufacturing industry. Many also believe that the environment is too demanding for women. While part of this may be true, have we ever stopped to consider what other reasons there might be?

Many women are well educated and they do find their way into their desired jobs, but on their way to the top they still tend to be excluded from what is referred to as "the old boys' club." On the other hand, I have met many progressive men who told me that they recognize the negative consequences on women — as well as other men — who don't want to participate in the exclusion of others through "the old boys' club." Other obstacles that may be preventing women from moving up are the lack of strong female sponsors or mentors, or they may have overlooked the importance of networking opportunities.

One of the biggest roadblocks for women in a technical, male-dominated environment is neither family nor children but rather the prejudices women have to deal with. Unfortunately, some women may become discouraged and with that, their self-worth and self-confidence may diminish. But it doesn't have to be that way. Women need to know what they are getting into and how to set their boundaries. It is important for them to express their expectations up front in order for their performance to be maximized and to ensure that their bosses are aware and supportive. This reduces frustration that may lead to the perception of inflexibility, lack of initiative or other issues.

Having lived most of my life in Austria, I don't believe that working for 16 hours a day necessarily means that you have been productive. It's not that Austrians don't work hard, but having a good work-life balance is important to them. I realize that we live in a fast-paced business environment and that we have to do more with fewer resources. However, it's equally important to properly organize and plan your days, weeks and months in order to get the most out of life. This might be particularly true for women, who still tend to be in charge of the children and home in addition to having careers.

You have to look after yourself, your health, your family and your well-being or you will eventually find yourself overwhelmed, unproductive and disillusioned in your job. This level of frustration will spill over to your family life and your overall level of happiness.

Why is it possible for most Austrians and Germans to go on vacation for three or four consecutive weeks while North Americans can't or don't, even if they have enough vacation time?

Of course, the problem also might be that yes, women too can think that no one in their company is as smart as they are, or that no one will be able to cover for them while they are away. Women and men with that worldview need to get over themselves. Everyone needs a vacation. The key is to allow someone to step in for you and you can return the favor. Not possible? Really? So what would happen if you had an accident or fell sick? Do you think life would go on? You bet it would! And consider that mental illnesses increase because of ongoing stress without proper rejuvenation time. Perhaps putting in long hours is an expectation or perhaps it's a matter of not being organized enough. What I do know for sure is that working 24/7 will burn you out.

Women are known as great multi-taskers, and life balance is important to them, especially if they have a family to care for. For employees who are in non-shift work positions or salaried leadership roles, the issue of travel and unplanned early or late meetings can be problematic. This can be especially true for women. Travel, off-hour meetings, dinners and events need to be planned more effectively to be more manageable. Overall, travel time is a barrier to women with children. Companies can benefit by investing in technology such as on-site or third-party video conferencing, or on-site Skype web-based technology on laptops, so that non-essential travel time and costs can be minimized.

You don't have to sign your life away when you sign a contract. Mental health issues, including depression and burnout syndromes for men and women, are increasing at an alarming rate. This is an issue that many organizations are grappling with. Healthcare costs are skyrocketing and solutions need to be found and implemented.

It is important for women to remain congruent with their values, and work with what is possible for them at their present stage in life. From an organizational standpoint, flexible hours, job sharing, affiliation with nearby daycares, or for larger entities on-site childcare, would demonstrate a sincere concern for employees' well-being. I know many companies see red when these things are mentioned but again, these programs need to be presented as a benefit to all employees, not exclusively to women, since in a majority of families both parents need to work. There is a significant benefit to employee performance and the bottom line when options such as these are available. When a man's daughter is ready to join the workforce, he begins to see the barriers through the eyes of his daughter. It is always a matter of perspective and if it affects you personally, you will tend to have a different outlook.

Do we encourage enough young women to choose engineering or science as a career? How do we prepare young women to be successful in the workforce? Another major reason that women have avoided manufacturing and engineering careers, rightly or wrongly, is their perception of the workplace. If they view a future career in manufacturing as a dirty job, with grueling hours, having to deal with a lack of respect by their male counterparts and difficulties moving up the ladder, of course they will feel less inclined to enter this industry.

Wouldn't women in leadership roles bring a certain element of sexiness to this industry? No, not in "that" way, gentlemen. **I sincerely believe that women could significantly make this industry more appealing. Women think, behave and act differently and that's what's needed.** We have to spark this industry, create more civilized and cultivated work environments, and establish an atmosphere wherein all people are able to envision futures for themselves and their loved ones. Let's stop for a moment, take a deep breath and consider what's missing. We have to actively search for different perspectives without becoming defensive. Together we can create a better tomorrow. Times are different and different actions will bring us different results.

I am not saying that female-dominated workplaces would change everything, but a good balance between men and women who respect each other's input will. The language will change and they will help each other see things that they wouldn't be able to see on their own.

I have met many women who are passionate about this industry and driven to succeed in their careers even though challenges such as managerial and industrial biases are certainly present. Even though women are faced with different obstacles and may have to work harder to get to where they want to be, with enough tenaciousness they will reach their goals. I believe that women's self-image is often not as good as their male counterparts'. Women question themselves about their ability to lead and succeed while most men are convinced that they are great and that they deserve the role they are in.

To all the women: Self-confidence and self-worth are key! There is a lot of pressure on you today. Don't get carried away by what people say you can or can't do. Simply think about the value you can and will bring to this industry. Don't look at what you need to have; look at what you have to give.

Let's consider the automotive industry. I was told that it is a "dirty" business and not intended for women. That is incorrect. According to Road & Travel Magazine, women purchase 65 percent of all new cars and 53 percent of used cars, and they influence 95 percent of all auto purchases. These are the facts. If women enjoy buying cars, why on earth wouldn't they want to be a part of designing, engineering and manufacturing them?

On the rare occasions that I meet a female General Manager or Production Manager, my first reaction is, "She must be really good to have gotten this far," and I bet I'm not alone. But I know that others might think something entirely different, which offers some supportive evidence to my contention that women have a tougher advancement path than do men in certain industries.

The environment in manufacturing plants is often rough and you have to be a tough cookie to be able to stand your ground. The current communication style leaves a lot of room for growth and in many companies the "F-word" is still widely used. It's shocking that nobody seems to think that there is anything wrong with this way of speaking to each other and it doesn't just happen on the shop floor. What happened to the simple lessons we learned in kindergarten? Manufacturing is a testosterone-driven industry. Unfortunately, the communication style of choice for many managers seems to be "Scream Therapy." The question is do women really want to put up with that? Should they have to? Should anyone? **There are enough words in the English language that people do not have to resort to rude, crude, unprofessional speech.**

Don't you think yelling and swearing are behaviors that have to be eliminated? As a matter of fact, I was told on more than one occasion that it is a sign of weakness if you employ more polite language. Is that why I notice bullies in managerial positions? Can it be an ingrained belief that bullies deliver better results than people who are capable of collaborating for a common goal? I cannot imagine any other reason why rude boors would be tolerated in the workplace. Through the enforcement of workplace violence and harassment legislation, this is starting to change. It doesn't matter whether you are male or female, kindness and respect have to return to the workplace.

Our corporations compete with each other for more and more. Profit is everything, regardless of the human toll. These corporations are led by people, and these individuals forget about everything that doesn't look like a financial statement — nature, animals, beauty, and more often than not, their own employees. What about coming from a place of equality and empathy? Can it be that this is really not possible or is it that people simply can't see the possibilities?

During one of my interviews, I asked a woman, "What can we do to attract more women to this industry?" She answered, "Women don't need to be attracted; they have to be accepted in leadership positions. Many times, jobs such as General Manager, Plant Manager and Production Manager are taboo."

This kind of belief can have a self-sabotaging effect and discourage women from following their hearts. If women accept a victim role, they become followers instead of leaders. Yes, it is tougher for them. Yes, there are still some backward thinkers at the top, but only through the sheer drive and tenacity of thick-skinned women determined to get on with their own success will this industry step up.

Negativity and frustration won't help women create the change that we would like to see. Women have to be strong advocates for fairness and support other women, men and diversity as a whole. Like everyone else in the organization, it is important that women accept the responsibility to be part of the solution. Blaming the problem on hierarchy or sexism and walking away is not the solution.

I believe the right attitude will help women to overcome any obstacle and with that, they will earn the respect of the men around them. Determination, persistence and action will help them to continue their journey so they can really go places and make a positive difference in the manufacturing world. **I believe the time for women in manufacturing has come. Can you hear us roar?**

Chapter 13

CAN WE MANUFACTURE PASSION?

**"Passion rebuilds the world for the youth.
It makes all things alive and significant."**

~ Ralph Waldo Emerson

I don't know why, but every time I walk into a manufacturing facility my heart starts to pound a little faster and my energy level soars. Perhaps it has something to do with my upbringing. My parents owned a construction business and I grew up with the good, the bad and the ugly of running your own business. It wasn't always easy for my parents, but I can remember the passion they both had for it. Like so many entrepreneurs, they started with a dream and, in their case, one big truck. Since I was a little girl, I learned to treat everyone with respect and my mom often said that our company was only as good as our people. When the employees were occasionally careless or damaged things, our reputation was on the line and I started to understand what it means to cultivate a workforce.

One lesson I learned from my parents early on was that if you want to go somewhere and there is no road, you can make a new one. It's as simple as that. Shouldn't this be true for any challenge we face in life? My inner drive to follow my dreams and my passion for encouraging organizations and individuals to follow theirs has led me on an amazing journey. I learn something new every day and each day I am faced with new challenges. I overcome negative self-talk and always search for new solutions and ideas. Sometimes my goals are overwhelming, but each day I gain valuable insights from the amazing people I surround myself with, and the people I meet.

It is always a thrill for me to observe the various manufacturing stages from the design and development to the finished product at the end of the line. High technology is the future but with it we face the danger of losing the human touch. Different departments and people have to collaborate with each other and their customers in order to make a desirable product. It's not just about making a profit. If we want to achieve operational excellence, emotional intelligence and a fervor for the future are key components.

Passionate leaders and employees can change the mindset, drive, dynamics and vision in any manufacturing plant. Without the zeal to create something bigger and better, there will be no change and companies won't be able to compete in the long term.

To compete and win with design and manufacturing worldwide is more than an exciting undertaking. It is sexy! If we allow employees to think outside the box and create a work environment where new and innovative ideas are rewarded, it can be incredibly inspiring for the whole team. It will also be incredibly sexy to create a longing in everyone to take part in that success. Can you imagine what would happen if you had the vision to create a product of the future and you encouraged your team to play a major role in producing it?

When was the last time you woke up and were excited about going to work? Are you passionate about what you do every day? Do you feel a sense of pride in the work that you do? Without it, it's very difficult to stay motivated. There are things we can learn and there are things we can't learn. Can we learn to be passionate? I have pondered this question many times. What do you think?

Since starting my business in January 2007, I have gone through some pretty difficult times. Who was going to hire an employee-engagement advisor when companies everywhere were laying off their employees? The recession was slowly starting to brew. Despite the setbacks, I weathered the storm and silenced the naysayers and doubters along the way. Why? Because I have an insatiable passion to make a positive impact in this industry and for me, failure is not an option.

People frequently told me to forget about manufacturing and focus on areas where I could make more money. They said I should concentrate on the banks, healthcare or the government. They didn't understand how committed I am to my vision of creating a better manufacturing industry. Their attitude was that trying to create change in this industry was hopeless. I believe when someone has the drive, energy and ambition to make things happen, anything is possible. When did people lose their spirit and stop following their hearts?

Many manufacturing work environments have employees with low energy. In such environments, people lack hope, but I see a glimmer of light. Let's start by asking ourselves how we will ever be able to successfully compete in a global marketplace when most people are on autopilot, just trying to survive the day.

Growing up in Austria, I can remember a quiz show on TV called "Made in Austria." This was a popular show and its purpose was to introduce consumers to high-quality products made in Austria. It established pride and awareness in viewers. Today we live in a very superficial world where most of us never take the time to consider where products such as food, clothing, toys, furniture and cars come from. Maybe we should take up a similar concept and create a TV show called "Made in North America" or "Made in Canada." After all, the United States saw the opportunity and created the "Made in America" brand, to which I say way to go!

The question is how can we reach a broader audience to teach them more about the challenges and opportunities in this industry? As we all know, many people watch TV and maybe we should consider better marketing for manufacturing on television. In Europe, TV entertainment includes one-hour roundtables with experts to explore different perspectives on various topics. My vision is to start a TV program with facilitated discussions. People from every walk of life who are passionate about this industry should be empowered to communicate their message on national television. Innovative companies, ideas or products could be featured and viewers could win a North American product by calling in and answering a question at the end of the program. Do you think this could be value added, informative and inspirational? **For me, there is no question that we have to include manufacturing in the entertainment industry because that's what will ultimately pique the general public's interest.**

There will always be people who make things possible and they are the ones with the power to change our world. There are also those who fight against what they don't know or understand and who resist change at all costs. It is up to us to ensure that these status-quo clingers don't destroy our spirit of innovation and creativity. There are many individuals who can't see the splendor of nature, the importance of preserving our environment, the potential of people, how their actions influence the big picture, or that it's possible to create something from nothing. I feel sorry for them.

Just as I am passionate about running marathons, I am passionate about doing whatever it takes — ethically, legally and morally, of course — to keep manufacturing in North America. So what does running a marathon have to do with running a successful business?

Running a marathon is not like sprinting. You have to start off slow and steady and have a strong mindset or you will soon run out of steam. Running a marathon requires working on getting better every single day. You have to have a goal and a plan, train to improve, surround yourself with supportive people, remain determined, and you better never forget to have fun along the way. If you blend passion into this mix, it makes it infinitely easier to follow through on what you want to accomplish. Don't you think the same holds true in business?

So many people seem to feel trapped, stressed, unfulfilled, depressed and even angry about the work they do or the people they have to deal with. Difficult economic times just add to the whole negative mindset, but it's time to stop blaming others for our frustrations and take a critical look at our own actions. Complacency, negativity and carelessness can be dominating factors in our workplaces. All too often, management has become so disconnected that they don't even notice the detrimental impact that this pervasive negative mindset has on productivity and profits.

I often wonder what the impact would be if we exchanged fanatical number crunchers with passionate leaders. Are business schools teaching everything there is to know about profit and loss and nothing about purpose and passion? Has ruthless greed taken all benevolence out of business? How can we allow the stock market to drive our outcomes rather than being mindful of the future we are creating for our children? Why don't we reward leaders for creating jobs instead of making cutthroat business decisions that won't serve anyone in the long run? Are we too focused on "having" to learn things rather than "wanting" to learn things?

Our world has become so complex and much of it is over-analyzed to the point where nothing gets done. Instead, let's consider ways that people can be truly engaged in their careers with a love for what they do. We need to encourage a society that values and respects every profession from street cleaner to doctor to toolmaker to rocket scientist. Skills have to be transferred through apprenticeship training and mentorship programs that place students with professionals from whom they can learn. Of course we can't just direct our employees: "Okay folks, get passionate about what you do." It doesn't work that way. You need to lead by example. Let your employees and co-workers see you doing your utmost to be the best you can be. Encourage them to be part of a winning team. Acknowledge and reward them for their contributions.

Ask yourself the following questions:

- **What can I do to combat complacency and foster an environment of creative and innovative thinking?**

- **Do I really appreciate my employees and regularly thank them for their great work?**

- **What's my attitude? Is it worth catching?**

- **How does my company respond to economic downturns?**

- **Do we continue to build for the long term even during difficult times by tapping into the creative solutions that employees have to offer?**

- **Does my company strive for excellence? Is our goal to do better tomorrow than we did today?**

- **Do my employees enjoy coming to work? Do they feel that they are part of the big picture?**

Can we manufacture passion? Yes, by doing the right things, we absolutely can. **Focus on improving yourself. If you work on becoming better at what you do and develop your mind and your character to become the best you can be, others will**

follow your lead. The actions of one person can create a ripple effect throughout an organization and change the entire work environment if it is supported at the top.

I wish that I could bottle my passion and positive energy and pass them along. There is no doubt that we can inspire people to see the big picture and get back to producing and purchasing North American products. **Passion is contagious and it is having a passion for what we do that makes a job sexy.**

Chapter 14

HAS MANUFACTURING LOST ITS MOJO FOR GENERATION Y?

"Children are the world's most valuable resource and its best hope for the future."
~ John F. Kennedy

How can we give manufacturing the recognition and attention it deserves? Have you ever wondered if we're doing everything possible to attract the next generation of skilled workers to the manufacturing industry? What are we doing to better understand what kind of work our youth is looking for?

Manufacturing is a skilled profession and many employers are concerned that people are not receiving the right education to equip them for jobs in this industry. **There is a tremendous shortage of skilled labor, which is often discussed but not often acted upon.** Educational programs exist, but how are they being promoted to fill up the classrooms? What can be done to get young people interested in apprenticeship programs if they are not aware that these opportunities exist?

Talent is the primary driver for manufacturing success. According to a 2011 report from Deloitte, manufacturers face a shortage of talent that is expected to worsen in the next three to five years. The majority of survey respondents also believe the talent shortage will increase in the coming years. They expect it to have the greatest impact on the workforce segments most critical to manufacturers. **Fully 67 percent of survey respondents face a shortage of qualified workers.** The shortage is most acute in critical segments such as skilled production workers, engineering technologists and scientists.

Industry leaders are a key component in encouraging youth to consider a career in manufacturing. More internship programs, more apprenticeship programs and more plant tours need to be offered to them at a younger age. The media also needs to play a role by casting a positive light on manufacturing by promoting the many great opportunities it has to offer; the opportunities to make a difference, to earn great wages and benefits, and enable long-term growth and viability.

It is time to wake up before it is too late. **There is a vast need for young people to get involved, but very few consider manufacturing careers. They are often unaware of the skills needed in an advanced manufacturing environment.** Professions such as welders, electricians, tool and die makers and machinists with specialized skills go far beyond pushing buttons or stacking boxes. Jobs include inspectors, quality assurance, programmers, lead hands, supervisors, machine mechanics, set-up technicians, sales personnel, robotic engineers, automation engineers, CAD designers, CAM engineers, mechanical engineers, and the list goes on and on and on.

If a business chooses to invest in and train young people, it will take it approximately three to five years to have someone trained to the level where he or she becomes **valuable to the company**. Don't you think that innovative products, which are also the most profitable products, are best produced by skilled and innovative employees?

Our educators, manufacturing associations, business owners and business leaders have to take the responsibility to educate the public with greater enthusiasm about apprenticeship programs and the significance of learning a trade. Parents naturally care about their children and want them to get into the right professions. Creating awareness is the first step in the right direction. **Our young men and young women have to experience the excitement, passion and pride that can come from producing a high-quality product in our home country.** Both parents and children have to see a bright future for this industry, which I doubt they can at the moment. Many parents who currently work in manufacturing don't see a future for their children there. If they can't see a future, who can?

In general, a job in manufacturing seems to have the same reputation as that of a garbage man. We all know that garbage collection is necessary, but it's not a job that we would persuade our own children to consider. Rather than being viewed as an important cornerstone to our wealth and prosperity, a career in manufacturing is looked upon as a possible option for those who are not smart enough to go to university. This may have been true in the past, but it no longer applies today because the trades truly are excellent career options.

Many children have fun building or repairing things. They enjoy being "hands-on," creating something from the simplest things. They have wonderful imaginations and are usually innately talented and creative. I am a big believer in encouraging people, young and old, to discover what they love to do. I would never suggest that someone should go into a profession just because the money is good. Interestingly, the majority of people don't know that employees in the manufacturing sector earn higher wages and receive more generous benefits than most other North American workers.

Young people who can combine national pride with a passion for excellence can and will do very well in this industry. Being part of something bigger can also engender excitement and motivation. Manufacturing is a viable profession. If we encourage the innovative and creative spirit in our children, what they design and produce in the future will change the world. We have great examples of people who did just that: Steve Jobs, Bill Gates, Sir Richard Branson and many others. The late Steve Jobs demanded that each new employee immediately become familiar with the Mac. After two weeks, these new employees were tested on it. Those who didn't pass were asked to leave the company. Those who passed received a Mac for their personal use. Jobs not only demanded excellence from his employees, he also expected them to be passionate about the quality of Apple's products. Not enough organizations share that mindset. Good enough is too often the order of the day.

We have to make sure that the employees in our organizations are connected to the end result. They have to be proud of the very important role they play in producing quality products.

How can we encourage our children to take an interest in and explore these career opportunities? One possibility would be to enter a co-op program where students can put the theory they have learned into practical application. Instead of accumulating debt, money can be earned while attending such a program. Once certification has been achieved, the job opportunities and financial rewards are exceptional.

Sometimes manufacturing may not be properly promoted. Remember "shop" class in high school, which mainly focused on males. It is also important to channel females' interest into manufacturing early in secondary school. How can we add a sexy element to the word "manufacturing" or even replace the word when it comes to attracting young women? The word is too broad and basic. It does not conjure up an exciting work environment for young women, or many young men, as they may associate it with repetitive, boring production lines and tough union stewards. I believe that by the seventh grade, schools owe all of their students a more positive, more fully encompassing picture of the manufacturing field than is currently being portrayed.

There is a tremendous need for greater awareness and understanding on the part of principals, teachers and school guidance counselors in order to connect the dots and work on improved cooperation between educators and manufacturers. There is such a lack of awareness by some educators that they don't know what they don't know, and yet they are advising our children on career choices!

Overall, I don't see that there is a good understanding of all the opportunities possible within the manufacturing sector. The whole concept of the supply chain must be better understood by young people, their parents and educators. Children need to be aware of the many career opportunities such as logistics, purchasing, procurement, science, environment, health and safety, human resources, business, finance, accounting and technology. Youth need to understand how interesting and fulfilling those careers can be. They also need to understand the positive impact they can have on the global economy.

There are now college and even Master's level degrees in supply chain management. It is a growing field that should be discussed and understood early in high school. Although guidance counselors will promote health, applied sciences and computer science as study options for university, I suspect that when they are discussing them, the careers are more in line with the standard professions such as medicine, engineering, research, quality laboratories, technology/software, designing and technical fields. However, these field descriptions may be far too narrow.

The assumption many people make is that manufacturing is laborious, monotonous and a dead-end job that requires only a basic level of education. Some people are still stuck in the early days of the industrial revolution and envision original assembly lines with dirty and unsafe work areas. Some also view manufacturing as being too labor intensive or low tech. Nothing could be further from the truth. A large segment of our population is devoted to this vital sector, and we can no longer afford to stand by and watch their talent, skills and abilities dwindle away.

Our youth want to work in environments where they can continuously improve and this requires basic skills. This approach makes a huge difference in the future success of any company. If potential employees feel that they are only being hired for their hands and not their brains, they won't see a future there. We need to dramatically change the outdated perceptions about manufacturing. The general public's perceptions about this sector are about 30 to 50 years out of date. We must start shifting associations of offshore sweatshops to modern operations with high-tech machinery and equipment, where people can create a quality product. This is the reality that is going to attract more and better candidates to manufacturing. We need to show that our industry can be interesting, challenging, motivating, and full of opportunities for growth.

How many teenagers graduate from high school today with the goal of working in a manufacturing plant? Not many, especially when their parents think that manufacturing is a dying industry in North America. How many parents would encourage their children to pursue a future in an uncertain area? Although many people work or have worked in this industry, it doesn't mean that the rest of the population fully understands the impact that a successful manufacturing sector can have on the future of their offspring.

Society today seems to favor academic education over skilled-trades training, and that's why many people who work in factories feel like second-class citizens. Why do we push our kids to go to college or university if they have an interest in and talent for learning a trade? There's a great German saying that comes to mind: **"A trade in hand finds gold in every land."**

I have come to realize that learning a skilled trade here doesn't carry the same value as it does in Austria, Switzerland or Germany.

Most parents here feel that their children have to go to university in order to make a good living. While this may have been true in the past, we don't need more lawyers or accountants. What we do need are skilled production workers! It is my experience that people in North America can earn a certificate for a specific profession in a relatively short period of time. Let's take Austria as an example. It takes three years to become a hairdresser, a salesperson in a department store, a tool and die maker, an electrician, and the list goes on and on. The learning process for these career options is common knowledge. If you ask someone in North America how long it takes to be fully trained as a hairdresser or salesperson, the majority of people wouldn't know. Austrians have a so-called dual apprenticeship. It is much like a co-op program that provides on-the-job-training in combination with a trade school for a period of three years. Students have the benefit of putting the theoretical learning into practical application and they are also financially compensated. After passing a final exam, they have acquired a skilled trade that is widely acknowledged and regulated by the Handcrafts Regulation Act.

As another example, an entry-level position at Volkswagen requires specific education and training. On its website, Volkswagen promotes the fact that it is willing to invest in people by providing them with the education and training they need to succeed at the company. Volkswagen's prerequisite is that it looks for young people who want to be the best at what they do and are able to demonstrate self-confidence and a team-oriented attitude. Volkswagen's unique approach to attracting and retaining outstanding employees is a key factor in its overall success.

The challenge is that Canadians don't spend enough on productivity and infrastructure. In Germany, companies incur the costs of their apprentices *before* they reap the benefits. Companies in North America should also be willing to invest in their future employees. Why wouldn't we consider implementing some of these best practices from other countries?

Many people believe that today's youth don't want to work hard. As I recall, the same thing was said about my generation when I was a teenager. Can it be that our youth can see the frustration their parents went through and have made a conscious decision that their lives will be different? They aspire for lives where they have more freedom and greater influence in the work that they do. They value the impossible and they are really interested in exploring and doing what hasn't been done before. Our young people are no better or worse than we used to be; they are just a different generation.

Management today has to understand that Generation Y needs to be managed differently. "Old-school" managers respect authority and hard work. They favor a top-down leadership style and take pride in a job well done. They expect seniority, experience and perseverance to be rewarded. I believe there needs to be mutual understanding and respect. The older generation needs to feel appreciated for its experience, knowledge and wisdom and the younger generation for its desire to create change, have work-life balance, and contribute to a cause.

The challenge will be that these Millennials may need more guidance and structure as they come from an overscheduled childhood. It will be up to us to challenge them to try and experience new things. Consider that they have been raised completely protected and they may expect work to continue this way. We have to demonstrate the value of standing out instead of fitting in and inspire them to take this industry to new heights.

There is still an attitude in companies that people should be grateful if they have a job. Employers feel that they are in a position of power, but this kind of power is already starting to vanish, especially since there are not enough skilled people to fill those jobs.

Our children are attracted to companies who are part of a cause, who want to do more than offer a service or product. Their attitude is about "hiring a boss" who treats them well and with respect, and is open and ready to acknowledge their ideas. Gen Y is not interested in a bottom line that shows only profit. They are interested in people and our planet. They believe in social responsibility. They want to go to work, be connected to the community and have a good time, and rightly so. They will work hard, but it won't only be work for the sake of work. Shouldn't this give us some hope for the future?

Today's youth is interested in high profile, attractive jobs and we haven't done a great job selling manufacturing in that context. The leading-edge technology used in manufacturing will certainly attract young people, but only if they are made aware of it. I said earlier that schools and universities play a critical role in working closely with manufacturing organizations to educate not only the students, but also the parents, teachers and career counselors about the many career opportunities this industry has to offer. Here, I offer some examples: field trips to manufacturing plants; manufacturing leaders going into schools to give presentations and answer any questions students may have; and co-op positions or internships in the manufacturing sector. As a manufacturer, it is your responsibility to go out and initiate these collaborations with academia. Some companies do this very well, but not enough manufacturers are being proactive.

Statistics show that only 30 percent of Americans said they have or would encourage their children to pursue a manufacturing career. Only 20 percent of parents have encouraged or will encourage their child or children to consider an engineering career. While American children and adults both feel that math and science are important, there is an ironic disconnect between recognizing the importance of and committing to pursuing a career in engineering and manufacturing.

The government also has a critical role to play in supporting our manufacturers, but as I mentioned before, we cannot just look to the government to create change because there are many things that we can start doing now. If organizations return to the basics of developing talent and listening to that talent, efficiency, effectiveness and profitability will be the end result.

When it comes to education, North American companies really have to adopt the European mindset of apprenticeship programs. The cost of university education is going up and often leaves young people with huge student debts. Oftentimes they graduate, but are unable to find jobs in their chosen fields.

More media coverage and TV shows would certainly help to promote manufacturing and other interesting ideas such as iPhone applications for manufacturing educational opportunities. Electronic games with manufacturing themes could also spark the interest of our younger population.

Jeremy Bout came up with a production called "The Edge Factor." This show really provides an inside look at how manufacturing impacts our lives, shapes our country, and is the backbone of our economy. I really enjoy this show! It is professionally done and in a manner that combines innovation and passion with the human aspect. One of Bout's goals with this show (www.edgefactor.com) is to demonstrate to our youth that manufacturing is incredibly diverse, interesting and, dare I say it, sexy.

There is another great TV program called "How It's Made." This show educates and enlightens people, especially our youth, about the many processes involved in creating a product. If you haven't seen it, you can check it out at http://science.discovery.com/tv/how-its-made/. A feature that could make this show even better would be to have the employees who work at these facilities share their stories of why they started to work there in the first place. I am sure we would be inspired by what we would hear.

The US has already initiated many things that will have a tremendous positive impact. The Student Summit is one excellent example. High school students in the US are invited to attend the International Manufacturing Technology Show in Chicago every year. Once high school students attend this show, they are hooked. Project Lead the Way (PLTW) is an American high school program that is making a difference in cultivating interest in children to do things with their hands and developing projects to engage them in manufacturing. There are also the Robotics and Battle-Bots competitions (www.battlebots.com) that are great. In fact, competitions are one of the most influential things being done right now to engage young people in manufacturing.

What are some first steps parents and children can take if they are interested in exploring a future in the industry? There are many things that can be done, but researching and exploring the main levels of interest will probably be the most challenging. If children learn about the many different opportunities available in manufacturing, they can become very excited about these career options. If children figure out why they want to work in the industry, it will be easy to guide them.

We are the subject matter experts. We are the teachers. We are the ones who must lead by example. We are the ones who need to be dissatisfied with the current status quo, take on the challenge and make improvements. We must create change. **This change should start with every single one of us, because if we don't get better, someone else will.** If we don't make manufacturing more attractive to today's youth, we will lose this potential labor pool to another sector or, even worse, to another country.

Outstanding resources:

http://www.manufacturingthefuture.com/

http://sme-tbm.org/tbm-roadmap-1/

http://www.imts.com/

http://www.pltw.org/

http://www.championnow.org/

http://www.manufacturingiscool.com/

http://www.setforjobs.org/

http://www.plantservices.com/skilltv/

http://skilltvtechnutia.blogspot.com/

http://www.antechsystems.com/occupyjobsapp/index.html

Chapter 15

COMMON SENSE: NOT AS COMMON AS YOU THINK

"Everybody gets so much information all day long that they lose their common sense."

~ Gertrude Stein

"I have no interest in becoming kinder. You can only get to the top if you are tough, lie and cheat. If you don't raise your voice, it is a sign of weakness. Karin, you have to understand that there are no business ethics in automotive."

This was a day in the fall of 2011 when someone tried to convince me that everything I believe in and everything I stand for is simply not practical in the "real world." I received this comment from a participant in one of my "Customer Obsession" workshops. The viewpoint of this individual did not change my beliefs or convictions. In fact, it had quite the opposite impact because his incendiary and erroneous comment made it here, into my book.

One can easily detect the level of stress and frustration expressed in his comment. My response to him was that his belief was actually very dangerous. He said, "That's not a belief. That's a fact."

I was hired by his employer because the management team felt something had to change regarding the way its employees were dealing with customers. I quickly discovered that their employees were on a vicious rollercoaster ride every day. Their behavior was reactive instead of proactive and they focused on problems instead of solutions. Obviously, this is an exhausting and depleting environment for anyone to have to endure day in and day out. In their dealings with their customers, they did the same things over and over again but expected a different result.

I do not begrudge the comment made by this fellow because over the years he has become disillusioned; he has no hope that anything is ever going to change and, honestly, without support from the management team, it won't. His level of discouragement shattered any dreams or aspirations that he could have a meaningful impact on the company's success. He is just one of many who feel defeated and who merely try to survive the day.

Have you noticed that mediocre managers work with mediocre employees? They are comfortable holding on to old ideas and behaviors because anything new would require a different mindset and a willingness to change. That's tough. That's taxing. This thought pushes them out of their comfort zones and so they continue doing what they have always done.

It was not surprising that the feedback in this workshop was mainly centered on unreasonable customer demands. Very little was said about striving for excellence, becoming better organized or producing a higher quality product to satisfy customer needs. It always, always, always comes down to money. If it's too expensive, they sacrifice quality and take the risk of an upset customer. Do we really believe this will lead to long-term success? I have heard similar comments before, but is this really how the majority of employees in manufacturing companies think? I certainly hope not.

When I step into this employee's shoes, I feel that he is right because he is only expressing what he sees on a daily basis. There really are too many people who are comfortable telling white lies to employees, co-workers, suppliers and customers. While this may be convenient, I think it really centers on how much of our own values we are willing to compromise. **It is astonishing to me that so few people have the determination and strength to stand up and say: "We can't do that. That is wrong. Let's find another way." Taking the high road may not be the easy thing to do, but it's the right thing to do.** As long as we keep rewarding people for doing the wrong thing to get more money in our pockets, these poisoned work environments will never change. That's one of the reasons that the level of disengagement is so high. These kinds of behaviors negatively impact people's productivity.

We should never assume engagement in the absence of evidence of disengagement. Unfortunately, disengaged employees don't show up at work with the word "disengaged" stamped on their foreheads. Sometimes everything on the surface may seem just fine, but the lack of engagement is often invisible and it can be damaging to team spirit and the overall success of the business.

Please take a moment and honestly answer the following questions:

- **Is your organization accepting or rejecting the status quo?**

- **What is not working for you right now?**

- **What are some of your biggest challenges when it comes to managing people?**

- **How do you think your bottom line is impacted when employees are negative, frustrated and disengaged?**

- **What are you doing right now to stay one step ahead of your competition?**

- **What role do your employees play with respect to innovation?**

- **Are you aware that your employees may be underused, misused and sometimes even abused?**

- **How willing are you to change and have a "change together with me" attitude?**

It has been said that "people don't trust you because they understand you; they trust you because you understand them."

Trust is so important — in relationships, in business, and in life in general. Many employees have a hard time trusting management. Some managers don't trust their employees. Investors have trust issues. Customers don't trust companies to provide the quality they were promised to receive. Suppliers don't trust that they will receive the money owed to them. What has happened? I wonder, doesn't anyone keep his or her word anymore?

- **Do you always tell the truth?**

- **Do you always keep your promises and commitments?**

- **Do you typically give your best effort and avoid cutting corners?**

- **Do you avoid using organizational resources for personal purposes?**

- **Do you stand up for what's right and take action to stop the misconduct by others?**

- **Do you do the easy thing or the right thing?**

If you are the boss, know that employees and co-workers are constantly watching you. They learn from you and will usually follow your lead. Ask yourself: "What lessons do I teach the people around me with respect to ethics?"

Although I do a lot of work in the automotive industry, I quite seriously feel that most managers and employees working in that industry and others live in a bubble. They have forgotten about the values they once held. In fact, someone recently told me that automotive is bad for one's character. Well, for mediocre companies, having good values combined with the drive to be the best in the industry is not the ultimate pursuit. Pushing the product out of the door is.

Good is good enough, and character and values are replaced by money and greed. I would say they live in their own self-absorbed world and it is quite devastating for me to see that there are no standards for character building, common-sense values, respectful behaviors, and positive "can-do" attitudes. Talk is cheap when actions speak louder than words!

I often think about common sense. As I spoke to people in various industries about this topic, I actually got the impression that many of them have given up on the notion that common sense is, in fact, common. Are people born with a certain potential to have or use common sense? And if they do have it, can they lose it?

First, we need to understand what common sense actually is. Common sense can be defined as ordinary, good sense; sound practical judgment that is independent of specialized knowledge, training or the like. Are most individuals today using or losing this important skill?

In one of my conversations, an individual brought something to my attention that has clearly stuck with me: Common sense is still in our minds, but it's losing its strength. So how can we bring back common sense and, with it, more drive and energy?

There was a time when companies made an effort to value this trait, and they took great pride in it. Out of the many comments I have received, this one in particular caught my attention: "Common sense isn't a learned behavior. It is either something that's in your DNA, or it isn't." I disagree as I would like to think that we all have common sense, but somehow, somewhere along the line, many of us have forgotten how to use it or were discouraged when we did use it. Sadly, I have observed that most people seem to think that today's work environment does not promote common sense.

We live in a society where many individuals feel that their opinions are the only ones that matter. As a result of this mentality, we have managed to create processes and definitions that are so complicated that others have little or no idea what they mean. Yet people are too proud to ask for clarification. This is one of many roadblocks to having common sense. The concept may be there, but few put it to use.

To get rid of confusion and gain perspective, we need to go back to the KISS theory: Keep It Short and Simple. Often, the logical answers are right under our noses, but people do not or choose not to see them. It takes a unique individual to actually do the right thing for the organization and to utilize common sense to solve a problem.

It is unfortunate that we have stopped living as a community and as a team, and have instead started living solely for ourselves. Instead of helping us become more independent, this mentality has actually made us even more sensitive to the actions of others. We have become so focused on building ourselves up that we have forgotten that our true foundation lies on things such as character and integrity.

In today's business world, we seem to be abandoning the tried and true "common-sense principles more often than not." Instead, we have reverted to the most basic human tendency: to choose immediate gratification over long-term stability; personal indulgence over public good; expediency over prudence. Common sense should tell us that these trends will eventually bring ruin to businesses and societies, as they have in the past. Still, it seems that we are losing the values required to exercise common sense. Is there a way to revitalize common sense thinking, and bring it back into the workplace?

I truly believe that with hard work and effort, common sense can once again be a prominent feature in any business. However, we need to recognize that what's common sense to me may not be common sense to you.

The ideas behind common sense do not just equate to having logic or an education. If all of us hold the same values, such as honesty, integrity, accountability, trust and compassion, and then stick by them, wouldn't that be a good start? Having common sense means standing by these values and if we bring them back into the business world, the results will be very progressive.

This is one of the many reasons that I am so passionate about working with people in a manufacturing environment. Most of them sincerely care about their jobs and want to do them well. If organizations rewarded employees for using this vital skill, it would result in better relationships, happier employees, and more successful businesses.

The art of common sense must filter its way through ego, human emotion and human want. Only then can it successfully be implemented in a business setting. For example, what holds management back from discussing certain problems with their shop floor employees? Promoting positive and open discussions that revolve around improvement opportunities is one way to re-introduce common sense in the workplace. Moreover, this will ultimately improve your business.

Toyota and Volkswagen are companies that lead the way and set high standards in this regard. It's a pleasure to follow the level of professionalism, organization and structure in their workplaces. I hope they will continue to hold their employees and their suppliers to the highest possible standards.

This year, I visited the Toyota plant in Cambridge, Ontario. It's difficult to describe the incredible positive energy I felt as I walked through that plant. They are so far ahead of the game. You may have heard about lean manufacturing. Lean is a production practice and management philosophy derived mostly from the Toyota Production System, also called TPS. TPS is embedded in Toyota's DNA and is built on 2 pillars: Kaizen, which is Japanese for continuous improvement, and Respect for People. Many books have been written about the Toyota Production System and yet very few companies are able to implement and sustain it.

The question is WHY?

Not long ago, a General Manager said to me, "Karin, you should use your approach in other countries because many manufacturing companies here, especially automotive, are still not warming up to the 'hugging,' as we say." What confuses me is that these same companies seem to have warmed up to lean manufacturing, but despite the investment they have made in training, they don't dedicate the same kind of resources to creating a culture of excellence.

Many companies focus too much on the specific lean tools rather than on their corporate culture, yet culture is the key to long-term sustainability and buy-in. Some management teams think they can delegate the lean responsibility to one or two individuals, but these people haven't grasped the whole philosophy. Having said this, unfortunately most North American companies don't get it. Here is my simple and unconventional insight: In order to create sustainable success with the lean production practice, it is fundamental to change your persistent habits of thought. A corporate culture can only start a sustainable transformation process when people begin focusing on themselves and learn self-governed behaviors. Consider that last year the employees of

Toyota Cambridge came up with 8,000 ideas and 90 percent of these ideas were implemented. Wow! The difference is that Toyota has the intent to implement as many ideas as possible because nobody knows better what to improve than the people who do the actual job. **That's Kaizen at its best: when you live it. This doesn't only make sense, it's common sense.**

Chapter 16

WHAT SHOULD MANUFACTURERS START DOING?

"You don't have to be great to start, but you have to start to be great."

~ Joe Sabah

Think!

Think before you act!

Think about a better way of doing things!

Think about how your behavior can affect others!

Think about what you are trying to do and why!

Think about whether your team is proactive or reactive!

Think about whether quality is more important than money!

Think about how new ideas are encouraged in your organization!

Think about where your best ideas come from!

Think about how you are able to sustain the idea flow!

Think about how free employees are to improve their work!

Think about what you want to be remembered for!

Individually and corporately, we need to become better at weighing the long-term consequences of our actions. You may not appreciate this statement at first, but I hope that as you read this chapter you will see the point I am trying to make. Based on what I have observed, people are so driven by a "go, go, go" or "getting the product out the door" mentality that they don't seem to have the time to think before they act. Many manufacturing companies are covered by a thick veil of problems, obstacles, roadblocks, stress, frustration and negativity. There isn't even enough time to rethink the actions that caused the quality issues or upset the customer in the first place. This is when I often hear people complain that the customer is too demanding.

What worked? What didn't work? How can we do better? How can we make better choices? What needs to change to produce different outcomes? These types of questions cannot be considered when the pressure is on and emotions are high. People react all day to situations fearing that if they don't react immediately, the world will come to an end. Feeling overwhelmed, underappreciated and anxious is very common. There is never enough time to do it right, but why do we always have the time to do it over again?

When we are reactive instead of proactive, we lose control. But is there anyone who wants to admit that he or she is not in control? Unfortunately, this vicious circle affects everyone else around us, too. If we don't know any better and if there is a lack of awareness, we simply cannot change this self-destructive behavior. That's when people start to blame others or cover up their mistakes whenever and wherever they can.

Just think about it: if there is never any resistance, healthy disagreement or conflict, isn't there a likelihood that nobody is thinking? Many people may have tried to give their input during crisis situations. However, when they get shut down too many times, it's easier for them to switch off brain function, wait for the paycheck and dial down to energy-saving mode until they go home. It happened to me when I first started working in the corporate world. I am an extrovert, which means I am outgoing and gregarious. In one of my jobs, I had to work in a cubicle all day long and I was told to be as quiet as possible. By the end of the business day, I felt brain dead. Sadly, I was led to believe that I had no choice other than to work in this manner.

I am sure the same has happened to many employees. They have lost hope — the hope that the work environment is going to change or get better. **They have become conditioned to allow others to "use" their brains for them.** We trust, we follow, we get frustrated, and then we complain. Henry Ford said, "Thinking is the hardest work there is, which is probably the reason why so few engage in it." Truer words have never been spoken.

What I have experienced is not that employees don't want to think, but rather that management hasn't discovered how to encourage and allow employees to tap into their own brainpower. **Employees should feel that they work in an environment where they can be creative and improve their processes no matter what their job functions are.** By not fostering an environment that encourages such initiatives, we won't be the best of the best. We won't be able to keep up the pace. This is one of the biggest differences between small and innovative mom-and-pop shops and mega-corporations. Start-ups and smaller companies have the drive to succeed and they actively seek input and imaginativeness. They look for their special niche and that's how they strive for excellence, perfection and profit. With that comes growth. This old saying may be true: "The smaller the niche, the faster you'll be rich."

With few exceptions, many large corporations seem to have lost the capacity to try new things, make mistakes, learn from them and move on. They have to develop the mindset of a start-up business. Mistakes can be costly and that's why companies would rather accept the status quo and settle for mediocrity.

Allow me to share a few personal experiences that demonstrate the danger of accepting the status quo and the arrogance that often accompanies it. Recently, an employee from a large organization said something to me that I can only describe as deeply shocking and lacking in sound business judgment. He said, "Our customer would be done if they didn't have us as a supplier." What a statement! This makes me think of my mom's favorite saying: "Pride comes before the fall."

Some time ago, I facilitated several workshops for a small group of managers. They recognized that their language toward the customer had to change, which is why they brought me on board. I conducted three workshops and suggested one-on-one coaching for the management team. Surprisingly, the General Manager did not see the value of coaching for his team because he would not consider it for himself. On the other hand, two of his driven, younger managers came to each workshop well prepared, only to experience the ridicule of the other, more seasoned team members, including the General Manager. Although these two managers took the workshops and assignments seriously, when the General Manager didn't, the entire project failed. Can you imagine the impact this had on the whole team? What a huge demotivator. Ironically, most of the managers complained that their customers treated them unfairly or with little respect, while telling me that they treated their suppliers the same way, if not worse.

Next is an experience I had with the very same company. I received payment and the check bounced. This had never happened to me before. It was not a huge sum of money, but what I was told when I spoke to someone from the accounting department was astounding. When I called them to tell them about the check, the person with whom I spoke said, "What's the big deal? Are you telling me you solely rely on our money? If that's the case, you might as well stop doing what you're doing and pack it in." I try not to take things personally, but for a moment I was speechless. This was the same company that had complained about its customers treating their people with utter disrespect, and now they were my customer. I guess the shoe was now on the other foot.

I remained calm and asked for a certified check. His frustrated response was, "Sure, but because of you I now have to go to the bank. You don't need a certified check. The next one will be okay." There was no apology and no intention to rectify the mistake quickly. Instead, he blamed me and made sure that as his supplier, I was aware of the inconvenience and trouble I had caused. I insisted on a certified check and eventually received it. What lesson did I learn? What goes around comes around! Moreover, if there are no clear expectations regarding how we ought to communicate with one another or with external customers and suppliers, incidents such as this will reoccur and negatively impact a company's reputation.

There may be times when a certain level of arrogance can be healthy, perhaps if you are part of a highly successful organization, but it needs to be tempered. Mediocre managers and employees believe that they have nothing left to learn. That's what's referred to as "destination disease," or fixed mindset. They think they have arrived. I wish them luck as they rest on their laurels and refuse to improve and progress. Managers who are resistant to change will be forced to accept an unpleasant truth: They will talk about innovation and creativity, but they won't be able to put it into action. They will want to improve quality, but they will not be able to connect with the people who can make it happen. They will not work on becoming better leaders because their main priority is getting their paychecks and bonuses.

I enjoy working with companies that want to be the best of the best, cultivate their teams, create awareness around character traits, and reconnect people with individual values and those of the company. It is crucial to create an environment of blue-sky thinkers. In order to make this happen, leadership has to be willing to be vulnerable. It is not easy to ask for and accept help.

Understanding how our daily choices impact everyone around us requires an investment of time and effort. It's not about completing a leadership course. **It is an ongoing process that allows us to acknowledge our own blind spots and shortcomings, talents and strengths. It is about becoming a better person, a better worker, a better leader and a better organization as a whole.**

Reflecting on my years in the corporate world, I know that I did many things during work hours that I probably shouldn't have done. I surfed the internet. I checked off a few things on my personal to-do list, and sometimes I lacked drive and initiative. Why? I was bored. I was not challenged enough and the majority of my managers were so busy with their own agendas that they simply didn't notice that I was capable of much more. I started to lose interest in giving my best. I don't think I was an exception. I just happen to be very honest. **For all the years I worked in the corporate world, I never reached a high level of creativity. When I think about those who worked with me, they didn't either. What a tremendous loss for those organizations!** It was only when I started my own business that I tapped into my ingenuity. I guess when you lack resources, you become resourceful. Why wouldn't the same be possible for you and your organization?

Don't you think it would be great if every single one of us would start challenging ourselves to consciously think about thinking consciously? Here is an example. Have you ever driven to work in the morning, but then you didn't even know how you got there? That's what I like to call the autopilot modus operandi. Many times throughout the day, you may be sucked into what you like to define as the same old problems and irritations. Nothing ever changes. This is a very dangerous state to be in because that's when our thoughts, habits and beliefs, combined with our assumptions, take over and allow us to make unconscious business decisions based on past experiences.

We cannot allow our teams to slip into a state of mental paralysis. This results in mediocrity in our thinking processes. **Being**

physically present at work doesn't necessarily mean people are mentally present. Attendance does not equal attention or productivity. From my personal experience, I know that working less is sometimes more. It's all about organizing and planning your day and accomplishing your tasks throughout the week to support your long-term goals.

Working within manufacturing environments, I hear firsthand that most shop floor employees feel like second-class citizens. Even though many management teams try hard to eradicate this way of thinking about their employees, their efforts don't seem to be successful in most cases. Why? The reason is simple: Their words do not align with their actions. I am convinced that they have the best intentions, but simply are unable to follow through. For example, if management's vision is to remove the barrier of "us versus them" and decisions are made behind closed doors without communicating the decisions to employees, employees won't feel included in the decision-making process. Too often I have heard responses like, "But Karin, we can't tell them everything. There are things that are none of their business." And in return they expect an environment of trust?

The importance of how managers communicate cannot be overemphasized. Their words have the power to either build employees up or tear them down. I have seen entire workforces being treated as just another tool to meet the bottom line with no regard to the fact that they are human beings with real feelings and emotions. In order to build a foundation of trust, absolutely everyone on the team, and that includes the boss, has to play by the same rules. Trust must be earned and earning it starts with management.

From an accounting perspective, people are shown as an expense on the profit and loss statement, while equipment and machinery are shown as an asset on the balance sheet. Can you see why we need a mental makeover? We have to remember that there is another reason for a business to exist, and that is to protect its most valuable assets: the people who produce what goes out the door and those who support those efforts. Awareness, education and celebration will make and keep manufacturing sexy! Ultimately, this will drive involvement, improvement and pride to be a part of something bigger and better.

Chapter 17

SOME THINGS JUST
DON'T ADD UP

**"This is your world. Shape it
or someone else will."**

~ Gary Lew

There are some things that just don't add up. **Sometimes I wonder what it would take to stop the things that go wrong in our corporate environments.** Organizations are led by people and if these people lack leadership skills, class or character, you will see it in their employees' faces, mindsets and productivity levels. **These companies may make money, but they fail to recognize how much they are missing by choosing not to see the opportunities.**

Unethical behaviors, corruption, liars and cheaters are rampant at the top of our organizations. Shouldn't we be concerned that authentic people with integrity are few and far between? How can we create a paradigm shift? The answer to this question is awareness.

Most managers are judged on quota and performance-based targets. They are focused on the bottom line. Most companies are purely data driven. **Of course, data is very important in being able to make sound business decisions, but if data is the only**

gauge by which managers are judged, there is a problem.
When those judgments dictate the probability of someone moving up in the company to greater levels of responsibility and
potentially higher pay, it's human nature to focus solely on what
will attain those targets. If a manager's bonus is primarily connected to the profitability of the company, do you think that
organizational improvements such as acquiring new equipment,
employee training, or enhancing the cafeteria will be on his or
her radar? How many managers would be willing to take a hit to
their bonus to do what they know they should do?

> People get rewarded for doing things right instead of doing
> the right things.
> People get rewarded for quarterly results instead of long-
> term viability.
> People get rewarded for saving costs instead of improving
> quality.
> People get rewarded for cutting jobs instead of saving or
> creating jobs.
> People get rewarded for giving answers instead of asking
> questions.

What's wrong with this picture? Am I perhaps too idealistic?
**Is it really only about the money or there is something of greater
importance at play?**
This reminds me of a good conversation that I recently had
with the General Manager of a large manufacturing company. He
told me that business was picking up and that they were hiring
again. This was good to hear; however, as the economy starts to
pick up, employees who may feel frustrated and disengaged in
their current jobs also start looking around for more satisfying
career opportunities. It's a double-edged sword. People will always
remember how they were treated during an economic downturn.
This General Manager mentioned that when he was asked to cut
employees, the direction that came from Head Office was to follow

the employees' contracts precisely. He was admonished not to pay out more than was required and to ensure that he completed an exit interview. He told me that he was disappointed that there was no mention of treating the people well or with dignity and respect.

The disappointment expressed by this General Manager shows that he sincerely cares about his employees. Head Office focused solely on the numbers. Memos such as this are sent by paper pushers, statistics junkies and data enthusiasts, who have no concern for people. Disconnection is one of most difficult challenges companies have to face. **I believe people are so concerned about the "what ifs" that they completely forget to focus on "how can we."**

What is your intention when you start the day? Are you determined to make it a good day and care about the people around you?

Let me tell you about an encounter I had with a janitor I spoke to after conducting one my workshops. I saw him beforehand and I noticed how happy and friendly he seemed to be. This particular evening, I had the chance to speak with him and I said, "You must really like working here because I always notice your upbeat attitude." He looked at me, smiled and replied, "Thanks for noticing. Well, I don't like everything that happens around here but if I complain about it, I will just be one more in a whole bunch of whiners and we have enough of those. That's why I try my best to set an example for others every day." He is a janitor and he chooses to be happy and he chooses to lead by example. Everyone can create change and everyone can be a leader. It's not a question of what position or title you hold. The game changer is how you view your position and how you consider your role and the impact you can make within the company. People start changing when they are able to look at their jobs from a different perspective.

In my workshops, I have noticed that many organizations lack trust. **Trust is fundamental — in relationships, in business and**

in life in general. Just compare it to building a house. A house needs a solid foundation to shoulder its considerable weight, provide a flat and level base for construction, and separate wood-based materials from contact with the ground, which would otherwise cause rot and allow for termite infestation.

We don't always see the significance of challenges in the simple and basic things that happen on a daily basis. We often don't appreciate the power and influence of personal behaviors such as respect and humility. **If we are constantly the sounding board for people who like to complain, we will never be able to really help them. If we make excuses for others, we are, in fact, discouraging them from changing.** We have to create greater awareness when it comes to our own actions and reactions with the people around us. It's simple: If you want people to change, you have to change and if you want things to get better, you have to become better and with that, the environment will change.

It really depends on how motivated we are to make daily continuous improvements both at home and at work. It's always easier to blame and see the faults of others. However, we have to remember that when we point the finger at someone else, three fingers are pointing back at us. That's not a very comforting thought, is it? **The best manufacturing organizations emphasize that self-discipline and continuous improvement are nothing less than a lifestyle!** It's the subtle discipline that occurs on an individual level and on a daily basis that will create change.

When I worked in Human Resources, I loved the diversity of the people I worked with. It was interesting and challenging. I recently read an article that said that in order to save money, many companies are opting for the CFO to take on HR responsibilities. I consider this a contradiction. Instead of CFOs, I would much rather have more visionaries in HR who see what employees can be. Don't you think it could be timely to rename Human Resources Managers and make them Human Relationship Managers? Isn't it discriminatory and de-valuing to see human beings as simple resources? In my opinion, a business will only be as good as the relationships it is able to build within and outside the organization.

When I started to work in HR at Magna, my boss suggested that I join Toastmasters. I did but for one reason only: to impress him. It may sound strange, but at the time I didn't know what I didn't know. I don't know if he recommended Toastmasters to anyone else in the department, so I figure it could have been for only one of two reasons. He either thought that because I was new to the country and a second-language speaker, I could benefit from the extra practice, or he saw the potential in me. Who knows, but I like to believe that it was the latter.

Have you ever attended a course or seminar not because you wanted to but because you had to? **Have you ever learned something that perhaps wasn't useful to you initially, but became priceless to you at a later time?** When I first started to attend Toastmasters meetings, I had no idea how much this learning experience would help me improve my communication, presentation and leadership skills. I am quite surprised that more people who speak English as their first language do not feel the need to join this organization. I am even more surprised that more organizations don't insist on establishing Corporate Toastmasters Groups at their workplaces. The art of getting a message across effectively is one of the most important skills one can have and still one of the hardest things to do. When I ask companies about some of their burning issues, communication is almost always at the top of the list.

Leaders know that communication is a problem. They know that ineffective communication holds them back from being more effective and yet many waste endless energy talking about it, instead of taking the necessary steps to making immediate improvements. How many managers would even remotely consider joining Toastmasters or working with a public-speaking coach? How many conferences have you attended where you had to listen to painfully tedious speakers? They think they shared their message but their audience stopped listening before they stopped talking. Have you ever considered how important your presentation skills are when you communicate? Outstanding presentation skills are critical in all walks of life and the higher you climb the ladder of success, the more important they become. If you are a better presenter, you will be better at selling yourself, you will be a better leader and, ultimately, you will be more successful in the long run.

Consider the following scenario: You are presenting at an employee meeting and your topic mostly consists of facts and figures. You see the eyes in your audience glazing over. People who are still alert are on their BlackBerrys or chuckling at the person beside them who has just dozed off. Ouch! The most amazing part is that most presenters just carry on with their presentation, ignore what's going on and just hope to plow through to the end. If presenters are really attentive, they may start to speak a little bit louder, but continue their message in the same monotonous tone. The real danger is when we think we have communicated but no one understood what we tried to say. I am not only talking about second-language speakers here.

As a manager, can you imagine the impact you could have on your employees if you would improve your communication skills and spruce up boring presentations by making them livelier and more interesting? How would you feel if people actually listened to what you had to say?

If you want to engage your audience, you have to change your presentation style. If you want the audience to pay more attention to you, you have to pay more attention to them. If you want them to be better listeners, you have to become a better speaker. It may sound simple, but it's not and it requires practice. Just imagine the potential in companies if more people would be willing to become better presenters and ultimately better communicators. As far as I am concerned, any corporation should be happy to have employees who are Toastmasters because these individuals value ongoing development and self-improvement. People who are better presenters, communicators and leaders will reduce one of the greatest time and money wasters: meetings! Companies need to start eliminating meeting madness.

Do you feel that you're attending too many pointless meetings? Are you falling behind in your real job because you have to rush from meeting to meeting? Do you ever tune out at meetings?

I believe that meetings are one of the most important ways for employees to communicate within an organization, but there are far too many that are too long, and most are boring and ineffective. Have you ever calculated the cost of ineffective meetings? It's astronomical. Here is my tip: Have fewer but better meetings.

I envision replacing the boardroom table with several treadmills. I am sure this will be the future of progressive companies. Every attendee would have to present his or her material while walking on the exercise equipment. People would not only be more creative and healthier; the cost savings would be incredible.

You don't see this happening any time soon? Well, you should! WARNING: **Sitting in meetings and sitting in cubicles without talking and being involved breeds stagnant minds.**

We want successful businesses that thrive and we worry about a disengaged workforce, but very few are willing to take a look at the root cause. I have the drive and desire to make a difference and since starting my business, I have met many wonderful people who make a difference each and every day. One of the disadvantages of working for a corporation is that you work in a bubble and you can lose sight of what's fundamental in the world. Are you happy with what you see when you look at yourself in the mirror at the end of the day?

During difficult economic times, the first thing that companies do to save costs is cut people and reduce improvement resources. This shows a tremendous lack of creativity and is highly unoriginal. It is my endeavor to create awareness that many people behind impressive desks or boardroom tables, or in small offices and cubicles, are sitting ducks. They are short-term thinkers and they are becoming as dull as the fluorescent lighting above them. It's the big picture that we all have to keep in mind. How does my behavior affect others? How can I respond differently to get a different outcome? How do my daily actions benefit the team? How will my decisions today affect my children tomorrow?

Many people believe that if their environment would change, their attitude would change as well. However, it is exactly the opposite. If you change your attitude, all the circumstances around will start to change. We have to understand that every single one of us plays an important role in the big picture. How do I make my buying decisions? Do I go out and vote? Do I stand up for what I believe in? Please don't be a person who doesn't bother to improve. Push yourself out of your comfort zone and engage your mind. Look for better ways of doing things and be humble enough to learn from the people around you. It will make you a better manager, a better employee and, what matters most, a better person. Your team is and always will be a reflection of you. At the end of the day, it is all about your attitude and your behaviors at home and in the workplace.

Let's try a little experiment. When you go to bed tonight, ask yourself the following question: "If my life were a movie, would I like the role I play in it?"

Chapter 18

VISION MEANS HAVING THE ABILITY TO PAINT A PICTURE

**"Sight has to do with what we can see,
but vision has to do with what we can be."**

~ Unknown

Are you tired of mediocrity? Are you annoyed with the status quo? Are you ready to create a movement and start a whole new dynamic? Would you like to join me in painting a picture of a better tomorrow?

Manufacturing should be front row and centre for the general public. It's the passion and excitement of making things that will draw more people to our industry. It's the desire to improve, to do better tomorrow than we did today. **It's the willingness to help each other become better at what we do without sacrificing ourselves and our values. There is a lot of power in working against the status quo.**

I have the deepest respect for people who make products that strengthen our communities and our economy. It takes courage and inspiration to manufacture our future. The next five years in manufacturing will be substantially different from anything we have ever seen before. For example, have you ever heard about "fabbing" or 3-D printing? This is an exciting technology that has been around for a while, but 3-D printing has been a fairly expensive process. As was the case with color printers, 3-D printers have been getting significantly cheaper over the past few years.

A "fabber" (short for "digital fabricator") is a factory in a box that makes things automatically from digital data. Fabbers generate three-dimensional, solid objects that you can hold in your hands, submit to testing, or assemble into working mechanisms. They are used by manufacturers around the world for low-volume production, prototyping and mold mastering. They are also used by scientists and surgeons for solid imaging, and by a few modern artists for innovative computerized sculptures. Manufacturers report enormous productivity gains through the use of fabbers. Such rapid prototyping allows designers to work faster and helps engineers catch problems before they reach production. Rapid prototyping has already started to morph into rapid, high-end manufacturing. We may find it difficult to grasp that software can be sent to a machine allowing it to print a complex component. In short, print-outs can be objects. Products that come out of 3-D printers are fully formed, and no assembly is required.

We will all have the ability to create things and to share our inventions with each other and the world. We can take the best ideas from each other and make truly great and awe-inspiring products. We will have manufacturing processes that encourage complexity and customization. In the future, we will have the opportunity to not only decide what our preferences of a product are but to participate in its actual design.

We should all be interested in reactivating our minds because the future will offer endless possibilities. Manufacturing will assuredly no longer be associated with repetitive jobs that offer opportunities for people with little or no formal education. We have to start changing this stereotype in our minds and start visualizing what's possible.

I have come to the realization that our strengths can sometimes also be our greatest weaknesses. This can be confusing at times, but that's when we have to take a step back, relax and reflect:

- **Who are we as an organization?**

- **Who should we be?**

- **How are we perceived?**

- **Where do we want to be?**

Please believe me when I say that it is critical to focus on doing the "greater good" for the manufacturing community, for our society, and for the future of our children. It has to be about more than just trying to "survive" the day. The number one question for every business owner should be, "How can we attract, fascinate and sustain our customers by aligning personal common-sense values with organizational values?"

In my eyes, a sexy industry embraces honesty, integrity, quality and respect in every single action taken both by management and by employees. If this seems impossible to you, that's okay. It's like running a race; an average athlete can't go at an elite runner's pace but he or she *can* make incremental improvements to get a little bit better every day.

We don't have to change everything; let's make improvements from where we are right now.

No matter how skilled or well informed we are, life will throw us a few curveballs. The question is, "How will we deal with them?" **We need to have the ability to bend but not break.**

I fully understand that everyone has to discover my message at their own pace. Those who know deep inside that there is some truth to what I have to say are those I want to reach. We have to create an environment with unique products and manufacture where we sell. **"Sexy manufacturing" is building the product in the region and nation in which it will be sold. This provides jobs for the manufacturing associates and enables them to be buyers of other products manufactured in their region and nation.** Henry Ford gave us this concept almost 100 years ago and the business model is still workable. Ford knew that he had to pay his workforce enough money that they were able to buy the cars they built. Simply put, manufacturing creates the goods that bring in the income that supports the service economy. I know many people who work two or three jobs in the service industry and they can barely make ends meet. We can't all just cut each other's hair and sell each other coffee. The income to pay for those haircuts and coffee has to come from somewhere.

Many big corporations overlook the innovation and creativity of smaller companies but it's these smaller companies that are eager to change our future and their resourcefulness is inspiring.

The key is to imagine and manufacture creative and innovative solutions because this is what customers are looking for and the workforce has to become a significant part of finding these innovative solutions. Proactive problem solvers and out-of-the-box thinkers will stand out rather than fit in. Wouldn't it be great to be rewarded for thinking "differently" instead of being shut down for new ideas we may have? We have to capitalize on things that other countries cannot do as well, such as new technologies and product development. I am well aware that there is a price to be paid to produce high-quality products. We have to invest time, money and energy, but isn't it worth the price to be known as "the best"?

We can sometimes get stuck with what we know, but to create a better tomorrow we have to be ready to venture into the unknown and to think about all of the possibilities out there.

If you want to be everything to everyone, you risk losing your focus and high standards of quality. There is tremendous power in the word "no" when we feel a client shouldn't be our client. On the other hand, if you want to become the best in the industry, you have to start with yourself as a leader and thus inspire your entire team to follow.

Do you want to know what I envision for the best of the best in a new manufacturing era?

I see leaders who:

- live by their values;
- respect others;
- value their character, attitude and behaviors;
- feel passionate and excited about making a product, and with that can strengthen the community and the economy;
- have high energy;
- are eager to take initiative;
- enjoy leading by example;
- are open minded enough to learn and to grow;
- don't talk down to others;
- see the value of every single person involved in the job;
- help others increase their own self-worth and confidence;
- treat their team as their customer;
- appreciate their customers and get excited about helping them solve their problems;
- want to become the best they can be;
- develop and support others to reach their full potential;
- have the courage to acknowledge their blind spots;
- have a "change together with me" attitude;

- ignite the brainpower of the people around them;

- value mentorship and sponsorship;

- make it part of their daily agenda to give back;

- start their careers on the production floor before climbing the corporate ladder, to gain an appreciation for the employees and the work they do;

- seek to understand others; and

- continue to remind themselves that their brain is no bigger than the brain of any other person in the company.

I see leaders who simply think more about how their actions and decisions influence our future. Take the time to reflect on the items listed above. **You don't have to change everything; start making small improvements from where you are right now. It's not about being perfect. It's about getting a little bit better each day.**

Please keep in mind that **I am not talking about managers, I am talking about leaders, and remember: You definitely don't need a title to be a leader!**

I envision:

- esthetically pleasing, proactive work environments with bright, colorful floors to inspire creativity;

- individuals working in different colored overalls;

- cutting-edge technology;

- a skilled and cultivated workforce;

- departments working together rather than against each other;

- meeting rooms with treadmills to keep meetings brief, efficient and effective;

- more work-life balance;

- greater emphasis on employee wellness with integrated gyms and physical activities being encouraged;

- new factories that look more like vacation resorts and where windows and natural light are mandatory;

- environmentally friendly work environments; and

- people at every level who want to do the right thing!

We have to jazz up manufacturing and show how the right atmospheres, environments and attitudes directly impact performance, production and employee satisfaction.

If you try to see what's possible, you can visualize many things. Once you do, it will eventually become a reality.

When I first started my business, I worked extensively with one of my leadership-coaching clients. In the middle of one of our sessions he said, "Karin, you are a dreamer. **Your ideas are great but they simply won't work in real life.**" I was stunned and did not know what to say, which is a rarity, believe me. Was I really an impractical dreamer? I started to ask a few people who know me well if they believed that I was a dreamer and I heard various answers. One said, "No, you are just very idealistic." Another said, "No, you sincerely care and you follow your heart wherever you go." A third said, "If you were a dreamer, you wouldn't put things into action." Finally, the last individual asked, "Karin, what's wrong with dreaming?" That is the response that has really stuck with me.

What *is* so wrong with dreaming? Did I feel insulted? Did I view this statement as something negative? Once again in my life, I was forced to change my perspective and I realized that the client who so astonished me had completely lost his ability to dream. Well, I was not about to let him destroy my dream.

Wouldn't you agree that this world is in great need of new ideas, new ways of doing things, new leaders, new inventions, new teaching methods, new ideals, new creations, and new ways of thinking and behaving? In reality, the world requires more dreamers who can and will put their dreams into action. After all, Martin Luther King once inspired millions with his speech, "I have a dream..."

If more leaders today would take advantage of the opportunity to inspire their employees with a clear vision for their organizations, I believe people would go above and beyond the call of duty. They will want to be part of a winning team and will help the company succeed on many levels. If we learn to combine information with inspiration, we will have mastered the art of motivational messaging.

Where would we be if we didn't have dreamers and pioneers who, despite the risks involved, took full advantage of every opportunity to realize their full potential? How would things be if people in the past had continued to do things the way they were always done? Perhaps we would still be crossing the ocean by ship rather than by plane.

Let us not forget, Thomas Edison dreamed of a lamp that could be operated by electricity... and invented it. Henry Ford, although poor and uneducated, dreamed of a "horseless carriage," and his dream changed the twentieth century.

During the past five years, I have had countless discouraging moments, but then I always look at a little post-it note on the wall right above my laptop: "Practical Dreamers DO NOT QUIT!" Not quitting has been a personal choice for me. How about you? Do you want to be a leader who searches for new meaning or will you choose to follow the crowd and avoid rocking the boat? We need more people with confidence — the confidence to speak up and the confidence to do the right thing. In my experience, dreams are not born of indifference, laziness or lack of ambition. Dreamers have an inner drive and they have a vivid picture before their eyes — pictures some call vision.

My most current unconventional insight is that dreaming really is a skill. It takes courage to dream and it requires even more skill to clearly and concisely communicate this dream to the world and to put it into action. Take advantage of the power of your imagination and start to think in pictures instead of numbers. The numbers will follow.

Does this sound too difficult or perhaps a little crazy? Just keep in mind that every child has boundless imagination and you were once a child, too. Somehow in the process of growing up, we start to lose this intrinsically valuable ability we once had.

- **How do you see your company in five years, in ten years?**

- **How many people will you have working with you?**

- **How will they approach challenges?**

- **What will the energy level in your work environment be?**

- **Will your company be micro-managed or will it be an idea factory?**

- **Will you want to implement more social technology?**

- **What will your customers say about you?**

- **What will your suppliers say about you?**

- **What will your employees say about you?**

- **What kind of a relationship will you have with your employees?**

On a personal level, I'd like you to ask yourself these questions:

- **Are you committed to buying quality North American products?**

- **Do you consider manufacturing to be a great career opportunity for your children?**

- **Are you ready to be proactive in a movement to manufacture a better future?**

As a final experiment, I want you to start learning to think in pictures. **Take some crayons and draw what your perfect day would look like.** It doesn't matter whether or not you can draw, just do it! Draw it, see it, feel it, and make it a reality! If you think it's juvenile or it makes you feel childish, all the better; we're trying to regain some of the magical thinking of childhood!

If you use the power of your imagination, you will really go places!

> *"You may say that I'm a dreamer*
> *but I'm not the only one.*
> *I hope one day you will join us*
> *and the world will live as one."*

John Lennon

I am proud to say that I am a dreamer. I can also say that I am in very good company. I hope that you too will join us.

Our economy must be built on a solid platform. We need to rebuild our infrastructure, renew our manufacturing base and educate our people. **If we put manufacturing at the forefront, we can employ our neighbors, our friends and our families to conceive, design, develop and produce the products we consume. This will strengthen our economy and for me, that is sexy!**

Afterword

"If you never change your mind, why have one?"
~ *Edward Bono*

As you read through this book, you may have wondered when I was going to provide you with some solutions to the issues I brought up. As I said in the preface, I consciously decided not to do that. I believe in working on solutions together.

I incorporated questions because asking as many questions as possible is the only way to get to the right answers. Every organization, every manager, every department and every employee has different challenges that call for different solutions. I sincerely believe that it makes sense to work on the best possible solution together and this should be free of complexity. Simplicity is key! You and your team have to feel out the best solutions for you in order to be fully committed to making them happen.

That said, here are my "Top 10 Common-Sense Insights" to make manufacturing "sexy." I wish you every success in becoming the best you can be. Have fun as you begin to experience exponential breakthroughs by engaging people's hearts and minds!

My Top 10 common-sense insights to make manufacturing sexy:

1. **Rediscover common-sense values and common etiquette; say "hello," "please," and "thank you!"**

2. **Develop possibility thinking; shift from a cost to quality mentality.**

3. Internalize a positive mindset and attitude; have the drive to make it a lifestyle of self-discipline and continuous improvement. Don't make excuses or get hung up on the little things.

4. Allow mistakes to happen and discover the creativity and innovation of your workforce.

5. Help employees understand how they contribute to the big picture and find ways to emotionally connect them to your business.

6. Reinforce Management by Walking Around (MBWA); get to know your employees. Tap into their unused brainpower and learn how to help them to reach their full potential.

7. Educate; develop an eagerness to learn and to grow within everyone in your organization. Discover the power of asking the right questions. Learn to listen and pay attention to the people who do the job every day.

8. Show respect; don't treat employees like children if you want them to behave like adults.

9. Shift from blame-storming meetings to brainstorming sessions and establish work and behavioral standards for your team.

10. Promote both internally and externally the fact that a skilled trade provides a golden foundation.

The proof of the pudding is in the eating !

Always remember that excellence is not a skill. It's an attitude. With the right attitude we can and will make manufacturing sexy!

Thank you for taking this journey with me. If you have any insights or ideas, I would love to hear from you. My email address is karin@karicosolutions.com.

Acknowledgements

**"At times our own light goes out and is
rekindled by a spark from another person.
Each of us has cause to think with deep
gratitude of those who have lighted
the flame within us."**

~ Albert Schweitzer

This book was only possible because of the collaboration of many people. The words of encouragement I have received from my family and friends during dark and discouraging moments and the insights and passion of wonderful people in the industry have helped me fulfill my dream of writing this book and conveying my heartfelt message to the world.

First and foremost, my former boss, my business associate, mentor, confidante and friend Heidi Garcia: Thank you for all your support, for the many hours spent editing this book, and for pushing and encouraging me to keep going. I wouldn't be where I am today without your friendship, loyalty and contribution.

My primary editor, Susan Chilton: Thank you for your dedication, commitment and hard work, and for making the editing process an exciting and insightful learning experience for me as an author. Your enthusiasm, brilliance and thoughtful comments made this book the best it could be.

My life partner Adrian Constantinescu: Thanks for your love, patience and support, and for being there for me during both the good and bad times. You are the one who brings out the best in me and makes me want to be a better person. It's you who helped me to "IronMind" my business!

My friend Phyllis Mancini, one of the smartest, most dedicated and most humble people I know: Thank you for acting as my editor. You are one special lady.

My friend Suzy Dickstein: Without you I wouldn't have had the courage to make How Can We Make Manufacturing Sexy the title of my book! Thank you.

My dear friends Marc Rapuch and Patricia Kay, who willingly listened to me throughout all of my ups and downs during this whole process: You both became my personal psychologists during our countless long runs. I truly appreciate our friendship and your continuous support.

My friend Alana Papeo: Thanks for all your encouragement and for helping me to become a better speaker and a better person.

My friend Marco Steinigans: You display all of the wonderful characteristics of a true friend. Thanks for your friendship and your continuous assistance and advice.

My friend Albert Egger: Although we may have had a rocky start, you are one of my most loyal supporters. Thanks for all your help.

My friend Mike Vickers: You keep reminding me that enjoying life to the fullest is the only way to be truly successful. Thanks for being my friend.

My friend Ferdinand Berger: Thanks for your support and always believing in me.

My friend Shelley O'Brien: Thanks for directing me to the exciting 3-D printing technology and for pointing out that some of this information should be included in this book.

My wonderful friends at Toastmasters International and especially my home club, Richmond Hill Toastmasters Club 1963: Thanks for making me a better communicator, speaker and leader. It's an honor to be a member of the Toastmasters community.

A special thank you to two phenomenal women: Professor Susan Helper at Case Western Reserve University and Krista Brookman at Deloitte who provided valuable assistance with research material.

A sincere "thank you" to all the people who shared their time and insights and gave me the opportunity to interview them: Bob Magee, President and CEO at The Woodbridge Group; Walter Palisca, CEO at Palcam Technologies; Rob Wildeboer, Executive Chairman at Martinrea International; Hemi Mitic, former Assistant to the President at CAW Canada; Katarina Gotovac, former VP of Operations at LHM Technologies; Bob Repovs, President and CEO at Samco Machinery; Vlad Minea, General Manager at McCowan Manufacturing; Alan N. Beaulieu, President at ETR Economics; Ralph Spaeth, General Manager at Theta TTS Precision Metal Forming; Rob Simpson, CEO at WSP Solutions; Daniel Richter, Founder and Director at Alpha Globe; Grant Vokey, Senior MES Architect and Project Lead at TCS; Bill Hennessey, CEO and Founder at Alio Industries; Lani Watson, President at Sustainable Lean Advantage; Terry Iverson, President at Iverson and Company; Richard Stape, Numerical Control Instructor at Youngstown State University; Zane Ferry, TPS/Lean Process Engineer at ADP Services; Harry Moser, President and Founder of the Reshoring Initiative; Jim Meehan, Managing Partner at Complete Manufacturing Group LLC; Sherman Yacher, Independent Electrical/Electronic Manufacturing Professional; and Frank Stronach, former Chairman at Magna International.

An earnest "Thank you" to Alfons Stadlbauer who helped me discover my inner artist with his amazing flipchart design workshops. He generously agreed to illustrate the various chapters in my book — all rights reserved. Please check out his website: www.alfons-stadlbauer.at/.

Kim Montefonte did a terrific job with the cover and text design. A heartfelt "thank you" for your creativity, your tremendous work ethic and your eagerness to make my book shine.

What would I have done without Heidy Lawrance? She expertly guided me through the self-publishing process. With her zest for excellence, Heidy and her outstanding team (www.wemake books.ca) helped me to design, produce and print my book not only within my expected time frame but also with high quality. Thank you for your patience and your support.

A huge "thank you" to all of the smart people on LinkedIn who continually share their knowledge and wisdom in the various group discussions. Without you, I wouldn't have had many of my insights.

Most importantly, a heartfelt "thank you" to my clients. You have taught me more than you will ever know.

To learn more about KARICO Performance
Solutions, it's individual and corporate
improvement approach, coaching services,
self-awareness workshops, and
inspirational speaking topics,
please call 1-647-401-5274 or email
info@karicosolutions.com.

Website
www.karicosolutions.com